Universitext

Universitext

Universitext is a series of textbooks that presents material from a wide variety of mathematical disciplines at master's level and beyond. The books, often well class-tested by their author, may have an informal, personal, even experimental approach to their subject matter. Some of the most successful and established books in the series have evolved through several editions, always following the evolution of teaching curricula, into very polished texts.

Thus as research topics trickle down into graduate-level teaching, first textbooks written for new, cutting-edge courses may make their way into *Universitext*.

More information about this series at http://www.springer.com/series/223

Ravindra B. Bapat

Graphs and Matrices

Second Edition

 HINDUSTAN
BOOK AGENCY

 Springer

Ravindra B. Bapat
Indian Statistical Institute
New Delhi
India

A co-publication with the Hindustan Book Agency, New Delhi, licensed for sale in all countries outside of India. Sold and distributed within India by the Hindustan Book Agency, P 19 Green Park Extn., New Delhi 110 016, India.
HBA ISBN 978-93-80250-66-3

ISSN 0172-5939   ISSN 2191-6675 (electronic)
ISBN 978-1-4471-6568-2 ISBN 978-1-4471-6569-9 (eBook)
DOI 10.1007/978-1-4471-6569-9

Library of Congress Control Number: 2014945463

Mathematics Subject Classification: 05C05, 05C12, 05C20, 05C38, 05C50, 05C57, 05C81, 15A09, 15A15, 15A18, 15B48

Springer London Heidelberg New York Dordrecht

Printed on acid-free paper

Springer is part of Springer Science+Business Media (www.springer.com)

Preface

This book is concerned with results in graph theory in which linear algebra and matrix theory play an important role. Although it is generally accepted that linear algebra can be an important component in the study of graphs, traditionally, graph theorists have remained by and large less than enthusiastic about using linear algebra. The results discussed here are usually treated under *algebraic graph theory*, as outlined in the classic books by Biggs [20] and by Godsil and Royle [39]. Our emphasis on matrix techniques is even greater than what is found in these and perhaps the subject matter discussed here might be termed *linear algebraic graph theory* to highlight this aspect.

After recalling some matrix preliminaries in the Chap. 1, the next few chapters outline the basic properties of some matrices associated with a graph. This is followed by topics in graph theory such as regular graphs and algebraic connectivity. Distance matrix of a tree and its generalized version for arbitrary graphs, the resistance matrix, are treated in the next two chapters. The final chapters treat other topics such as the Laplacian eigenvalues of threshold graphs, the positive definite completion problem, and matrix games based on a graph.

We have kept the treatment at a fairly elementary level and resisted the temptation of presenting up-to-date research work. Thus, several chapters in this book may be viewed as an invitation to a vast area of vigorous current research. Only a beginning is made here with the hope that it will entice the reader to explore further. In the same vein, we often do not present the results in their full generality, but present only a simpler version that captures the elegance of the result. Weighted graphs are avoided, although most results presented here have weighted, and hence more general, analogs.

The references for each chapter are listed at the end of the chapter. In addition, a master bibliography is included. In a short note at the end of each chapter, we indicate the primary references that we used. Often, we have given a different treatment, as well as different proofs, of the results cited. We do not go into an elaborate description of such differences.

It is a pleasure to thank Rajendra Bhatia for his diligent handling of the manuscript. Aloke Dey, Arbind Lal, Sukanta Pati, Sharad Sane, S. Sivaramakrishnan, and Murali Srinivasan read either all or parts of the manuscript, suggested changes and pointed out corrections. I sincerely thank them all. Thanks are also due to the anonymous referees for helpful comments. Needless to say I remain responsible for the shortcomings and errors that persist. The facilities provided by the Indian Statistical Institute, New Delhi, and the support of the JC Bose Fellowship, Department of Science and Technology, Government of India, are gratefully acknowledged.

New Delhi, India   Ravindra B. Bapat

About the Second Edition

In this edition, besides correcting some errors and typos in the first edition, we have added a new chapter on the line graph of a tree.

I sincerely thank Nazli Besharati, Arbind K. Lal, Ambat Vijayakumar, Anu Varghese, and Seethu Varghese for pointing out corrections in the first edition. I also thank Souvik Dhara, Ibrahim Ghorbani, and Rajesh Kannan for a careful reading of the new chapter and for helpful suggestions.

Contents

Chapter 1
Preliminaries

In this chapter we review certain basic concepts from linear algebra. We consider only real matrices. Although our treatment is self-contained, the reader is assumed to be familiar with the basic operations on matrices. Relevant concepts and results are given, although we omit proofs.

1.1 Matrices

Basic Definitions

An $m \times n$ matrix consists of mn real numbers arranged in m rows and n columns. The entry in row i and column j of the matrix A is denoted by a_{ij}. An $m \times 1$ matrix is called a column vector of order m; similarly, a $1 \times n$ matrix is a row vector of order n. An $m \times n$ matrix is called a square matrix if $m = n$.

Operations of matrix addition, scalar multiplication and matrix multiplication are basic and will not be recalled here. The transpose of the $m \times n$ matrix A is denoted by A'.

A diagonal matrix is a square matrix A such that $a_{ij} = 0, i \neq j$. We denote the diagonal matrix

$$
\begin{bmatrix}
\lambda_1 & 0 & \cdots & 0 \\
0 & \lambda_2 & \cdots & 0 \\
\vdots & \vdots & \ddots & \vdots \\
0 & 0 & \cdots & \lambda_n
\end{bmatrix}
$$

by $\mathsf{diag}(\lambda_1, \ldots, \lambda_n)$. When $\lambda_i = 1$ for all i, this matrix reduces to the identity matrix of order n, which we denote by I_n or often simply by I if the order is clear from the context. The matrix A is upper triangular if $a_{ij} = 0, i > j$. The transpose of an upper triangular matrix is lower triangular.

© Springer-Verlag London 2014
R.B. Bapat, *Graphs and Matrices*, Universitext,
DOI 10.1007/978-1-4471-6569-9_1

Trace and Determinant

Let A be a square matrix of order n. The entries $a_{11}, \ldots, a_{nn}$ are said to constitute the (main) diagonal of A. The *trace* of A is defined as

$$\operatorname{trace} A = a_{11} + \cdots + a_{nn}.$$

It follows from this definition that if A, B are matrices such that both AB and BA are defined, then

$$\operatorname{trace} AB = \operatorname{trace} BA.$$

The *determinant* of an $n \times n$ matrix A, denoted by $\det A$, is defined as

$$\det A = \sum_{\sigma} sgn(\sigma) a_{1\sigma(1)} \cdots a_{n\sigma(n)},$$

where the summation is over all permutations $\sigma(1), \ldots, \sigma(n)$ of $1, \ldots, n$, and $sgn(\sigma)$ is 1 or -1 according as σ is even or odd. We assume familiarity with the basic properties of determinant.

Vector Spaces Associated with a Matrix

Let $\mathbb{R}$ denote the set of real numbers. Consider the set of all column vectors of order n ($n \times 1$ matrices) and the set of all row vectors of order n ($1 \times n$ matrices). Both of these sets will be denoted by $\mathbb{R}^n$. We will write the elements of $\mathbb{R}^n$ either as column vectors or as row vectors, depending upon whichever is convenient in a given situation. Recall that $\mathbb{R}^n$ is a vector space with the operations matrix addition and scalar multiplication.

Let A be an $m \times n$ matrix. The subspace of $\mathbb{R}^m$ spanned by the column vectors of A is called the *column space* or the *column span* of A. Similarly the subspace of $\mathbb{R}^n$ spanned by the row vectors of A is called the row space of A.

According to the fundamental theorem of linear algebra, the dimension of the column space of a matrix equals the dimension of the row space, and the common value is called the *rank* of the matrix. We denote the rank of the matrix A by $\operatorname{rank} A$.

For any matrix A, $\operatorname{rank} A = \operatorname{rank} A'$. If A and B are matrices of the same order, then $\operatorname{rank}(A + B) \leq \operatorname{rank} A + \operatorname{rank} B$. If A and B are matrices such that AB is defined, then $\operatorname{rank} AB \leq \min\{\operatorname{rank} A, \operatorname{rank} B\}$.

Let A be an $m \times n$ matrix. The set of all vectors $x \in \mathbb{R}^n$ such that $Ax = 0$ is easily seen to be a subspace of $\mathbb{R}^n$. This subspace is called the *null space* of A, and we denote it by $\mathscr{N}(A)$. The dimension of $\mathscr{N}(A)$ is called the *nullity* of A. Let A be an $m \times n$ matrix. Then the nullity of A equals $n - \operatorname{rank} A$.

Minors

Let A be an $m \times n$ matrix. If $S \subset \{1, \ldots, m\}$, $T \subset \{1, \ldots, n\}$, then $A[S|T]$ will denote the submatrix of A determined by the rows corresponding to S and the columns corresponding to T. The submatrix obtained by deleting the rows in S and the columns in T will be denoted by $A(S|T)$. Thus, $A(S|T) = A[S^c|T^c]$, where the superscript c denotes complement. Often, we tacitly assume that S and T are such that these matrices are not vacuous. When $S = \{i\}$, $T = \{j\}$ are singletons, then $A(S|T)$ is denoted $A(i|j)$.

Nonsingular Matrices

A matrix A of order $n \times n$ is said to be *nonsingular* if rank $A = n$; otherwise the matrix is *singular*. If A is nonsingular, then there is a unique $n \times n$ matrix A^{-1}, called the inverse of A, such that $AA^{-1} = A^{-1}A = I$. A matrix is nonsingular if and only if det A is nonzero.

The cofactor of a_{ij} is defined as $(-1)^{i+j}$ det $A(i|j)$. The adjoint of A is the $n \times n$ matrix whose (i, j)th entry is the cofactor of a_{ji}. We recall that if A is nonsingular, then A^{-1} is given by $\dfrac{1}{\det A}$ times the adjoint of A.

A matrix is said to have full column rank if its rank equals the number of columns, or equivalently, the columns are linearly independent. Similarly, a matrix has full row rank if its rows are linearly independent. If B has full column rank, then it admits a left inverse, that is, a matrix X such that $XB = I$. Similarly, if C has full row rank, then it has a right inverse, that is, a matrix Y such that $CY = I$.

If A is an $m \times n$ matrix of rank r then we can write $A = BC$, where B is $m \times r$ of full column rank and C is $r \times n$ of full row rank. This is called a *rank factorization* of A. There exist nonsingular matrices P and Q of order $m \times m$ and $n \times n$, respectively, such that

$$A = P \begin{bmatrix} I_r & 0 \\ 0 & 0 \end{bmatrix} Q.$$

This is the rank canonical form of A.

Orthogonality

Vectors x, y in $\mathbb{R}^n$ are said to be orthogonal, or perpendicular, if $x'y = 0$. A set of vectors $\{x_1, \ldots, x_m\}$ in $\mathbb{R}^n$ is said to form an *orthonormal basis* for the vector space S if the set is a basis for S, and furthermore $x_i'x_j$ is 0 if $i \neq j$, and 1 if $i = j$. The $n \times n$ matrix P is said to be orthogonal if $PP' = P'P = I$. One can verify that if P is orthogonal then P' is orthogonal.

If $x_1, \ldots, x_k$ are linearly independent vectors then by the Gram–Schmidt orthogonalization process we may construct orthonormal vectors $y_1, \ldots, y_k$ such that y_i is a linear combination of $x_1, \ldots, x_i$; $i = 1, \ldots, k$.

Schur Complement

Let A be an $n \times n$ matrix partitioned as

$$A = \begin{bmatrix} A_{11} & A_{12} \\ A_{21} & A_{22} \end{bmatrix}, \tag{1.1}$$

where A_{11} and A_{22} are square matrices. If A_{11} is nonsingular then the *Schur complement* of A_{11} in A is defined to be the matrix $A_{22} - A_{21}A_{11}^{-1}A_{12}$. Similarly, if A_{22} is nonsingular then the Schur complement of A_{22} in A is $A_{11} - A_{12}A_{22}^{-1}A_{21}$.

The following identity is easily verified:

$$\begin{bmatrix} I & 0 \\ -A_{21}A_{11}^{-1} & I \end{bmatrix} \begin{bmatrix} A_{11} & A_{12} \\ A_{21} & A_{22} \end{bmatrix} \begin{bmatrix} I & -A_{11}^{-1}A_{12} \\ 0 & I \end{bmatrix} = \begin{bmatrix} A_{11} & 0 \\ 0 & A_{22} - A_{21}A_{11}^{-1}A_{12} \end{bmatrix}. \tag{1.2}$$

The following useful fact can be easily proved using (1.2):

$$\det A = (\det A_{11}) \det(A_{22} - A_{21}A_{11}^{-1}A_{12}). \tag{1.3}$$

We will refer to (1.3) as the Schur complement formula, or the Schur formula, for the determinant.

Inverse of a Partitioned Matrix

Let A be an $n \times n$ nonsingular matrix partitioned as in (1.1). Suppose A_{11} is square and nonsingular and let $A/A_{11} = A_{22} - A_{21}A_{11}^{-1}A_{12}$ be the Schur complement of A_{11}. Then

$$A^{-1} = \begin{bmatrix} A_{11}^{-1} + A_{11}^{-1}A_{12}(A/A_{11})^{-1}A_{21}A_{11}^{-1} & -A_{11}^{-1}A_{12}(A/A_{11})^{-1} \\ -(A/A_{11})^{-1}A_{21}A_{11}^{-1} & (A/A_{11})^{-1} \end{bmatrix}.$$

Note that if A and A_{11} are nonsingular, then A/A_{11} must be nonsingular. Equivalent formulae may be given in terms of the Schur complement of A_{22}.

Cauchy–Binet Formula

Let A and B be matrices of order $m \times n$ and $n \times m$ respectively, where $m \leq n$. Then

$$\det(AB) = \sum \det A[\{1, \ldots, m\}|S] \det B[S|\{1, \ldots, m\}],$$

where the summation is over all m-element subsets of $\{1, \ldots, n\}$.
  To illustrate by an example, let

$$A = \begin{bmatrix} 2 & 3 & -1 \\ 4 & 0 & 2 \end{bmatrix}, \quad B = \begin{bmatrix} 1 & -2 \\ 0 & 3 \\ 5 & 1 \end{bmatrix}.$$

Then $\det(AB)$ equals

$$\det \begin{bmatrix} 2 & 3 \\ 4 & 0 \end{bmatrix} \begin{bmatrix} 1 & -2 \\ 0 & 3 \end{bmatrix} + \det \begin{bmatrix} 2 & -1 \\ 4 & 2 \end{bmatrix} \begin{bmatrix} 1 & -2 \\ 5 & 1 \end{bmatrix} + \det \begin{bmatrix} 3 & -1 \\ 0 & 2 \end{bmatrix} \begin{bmatrix} 0 & 3 \\ 5 & 1 \end{bmatrix}.$$

1.2 Eigenvalues of Symmetric Matrices

Characteristic Polynomial

Let A be an $n \times n$ matrix. The determinant $\det(A - \lambda I)$ is a polynomial in the (complex) variable λ of degree n and is called the *characteristic polynomial* of A. The equation

$$\det(A - \lambda I) = 0$$

is called the *characteristic equation* of A. By the fundamental theorem of algebra the equation has n complex roots and these roots are called the *eigenvalues* of A.
  We remark that it is customary to define the characteristic polynomial of A as $\det(\lambda I - A)$ as well. This does not affect the eigenvalues.
  The eigenvalues might not all be distinct. The number of times an eigenvalue occurs as a root of the characteristic equation is called the *algebraic multiplicity* of the eigenvalue.
  We may factor the characteristic polynomial as

$$\det(A - \lambda I) = (\lambda_1 - \lambda) \cdots (\lambda_n - \lambda).$$

  The geometric multiplicity of the eigenvalue λ of A is defined to be the dimension of the null space of $A - \lambda I$. The geometric multiplicity of an eigenvalue does not exceed its algebraic multiplicity.

If A and B are matrices of order $m \times n$ and $n \times m$, respectively, where $m \geq n$, then the eigenvalues of AB are the same as the eigenvalues of BA, along with 0 with a (possibly further) multiplicity of $m - n$.

If $\lambda_1, \ldots, \lambda_n$ are the eigenvalues of A, then $\det A = \lambda_1 \cdots \lambda_n$, while $trace\ A = \lambda_1 + \cdots + \lambda_n$.

A principal submatrix of a square matrix is a submatrix formed by a set of rows and the corresponding set of columns. A principal minor of A is the determinant of a principal submatrix. A leading principal minor is a principal minor involving rows and columns $1, \ldots, k$ for some k.

The sum of the products of the eigenvalues, of A, taken k at a time, equals the sum of the $k \times k$ principal minors of A. When $k = 1$ this reduces to the familiar fact that the sum of the eigenvalues equals the trace.

If $\lambda_1, \ldots, \lambda_n$ are the eigenvalues of the $n \times n$ matrix A, and if $q(A)$ is a polynomial in A, then the eigenvalues of $q(A)$ are $q(\lambda_1), \ldots, q(\lambda_n)$.

If A is an $n \times n$ matrix with the characteristic polynomial $p(A)$, then the Cayley–Hamilton theorem asserts that $p(A) = 0$. The monic polynomial $q(x)$ of minimum degree that satisfies $q(A) = 0$ is called the *minimal polynomial* of A.

Spectral Theorem

A square matrix A is called *symmetric* if $A = A'$. The eigenvalues of a symmetric matrix are real. Furthermore, if A is a symmetric $n \times n$ matrix, then according to the spectral theorem there exists an orthogonal matrix P such that

$$PAP' = \begin{bmatrix} \lambda_1 & 0 & \cdots & 0 \\ 0 & \lambda_2 & \cdots & 0 \\ \vdots & \vdots & \ddots & \vdots \\ 0 & 0 & \cdots & \lambda_n \end{bmatrix}.$$

In the case of a symmetric matrix the algebraic and the geometric multiplicities of any eigenvalue coincide. Also, the rank of the matrix equals the number of nonzero eigenvalues, counting multiplicities.

Let A and B be symmetric $n \times n$ matrices such that they commute, i.e., $AB = BA$. Then A and B can be simultaneously diagonalized, that is, there exists an orthogonal matrix P such that PAP' and PBP' are both diagonal, with the eigenvalues of A (respectively, B) along the diagonal PAP' (respectively, PBP').

Positive Definite Matrices

An $n \times n$ matrix A is said to be *positive definite* if it is symmetric and if for any nonzero vector x, $x'Ax > 0$. The identity matrix is clearly positive definite and so is a diagonal matrix with only positive entries along the diagonal. Let A be a symmetric $n \times n$ matrix. Then any of the following conditions is equivalent to A being positive definite:

(i) the eigenvalues of A are positive;
(ii) all principal minors of A are positive;
(iii) all leading principal minors of A are positive;
(iv) $A = BB'$ for some matrix B of full column rank;
(v) $A = TT'$ for some lower triangular matrix T with positive diagonal entries.

A symmetric matrix A is called *positive semidefinite* if $x'Ax \geq 0$ for any x. Equivalent conditions for a matrix to be positive semidefinite can be given similarly. However, note that the leading principal minors of A may be nonnegative and yet A may not be positive semidefinite. This is illustrated by the example $\begin{bmatrix} 0 & 0 \\ 0 & -1 \end{bmatrix}$. Also, in (v), the diagonal entries of T need only be nonnegative.

If A is positive semidefinite then there exists a unique positive semidefinite matrix B such that $B^2 = A$. The matrix B is called the *square root* of A and is denoted by $A^{1/2}$.

Let A be an $n \times n$ matrix partitioned as

$$A = \begin{bmatrix} A_{11} & A_{12} \\ A_{21} & A_{22} \end{bmatrix}, \tag{1.4}$$

where A_{11} and A_{22} are square matrices.

The following facts can be easily proved using (1.2):

(i) If A is positive definite then $A_{22} - A_{21}A_{11}^{-1}A_{12}$ is positive definite;
(ii) Let A be symmetric. If A_{11} and its Schur complement $A_{22} - A_{21}A_{11}^{-1}A_{12}$ are both positive definite then A is positive definite.

Interlacing for Eigenvalues

The following result, known as the Cauchy interlacing theorem, finds considerable use in graph theory.

Let A be a symmetric $n \times n$ matrix and let B be a principal submatrix of A of order $n - 1$. If $\lambda_1 \geq \cdots \geq \lambda_n$ and $\mu_1 \geq \cdots \geq \mu_{n-1}$ are the eigenvalues of A and B, respectively, then

$$\lambda_1 \geq \mu_1 \geq \lambda_2 \geq \cdots \geq \lambda_{n-1} \geq \mu_{n-1} \geq \lambda_n. \tag{1.5}$$

A related interlacing result is as follows. Let A and B be symmetric $n \times n$ matrices and let $A = B + xx'$ for some vector x. If $\lambda_1 \geq \cdots \geq \lambda_n$ and $\mu_1 \geq \cdots \geq \mu_n$ are the eigenvalues of A and B respectively, then

$$\lambda_1 \geq \mu_1 \geq \lambda_2 \geq \cdots \geq \lambda_n \geq \mu_n. \tag{1.6}$$

Let A be a symmetric $n \times n$ matrix with eigenvalues $\lambda_1(A) \geq \cdots \geq \lambda_n(A)$, arranged in nonincreasing order. Let $||x||$ denote the usual Euclidean norm, $\left(\sum_{i=1}^n x_i^2\right)^{\frac{1}{2}}$. The following extremal representation will be useful:

$$\lambda_1(A) = \max_{||x||=1}\{x'Ax\}, \quad \lambda_n(A) = \min_{||x||=1}\{x'Ax\}.$$

Setting x to be the ith column of I in the above representation we see that

$$\lambda_n(A) \leq \min_i\{a_{ii}\} \leq \max_i\{a_{ii}\} \leq \lambda_1(A).$$

1.3 Generalized Inverses

Let A be an $m \times n$ matrix. A matrix G of order $n \times m$ is said to be a *generalized inverse* (or a g-*inverse*) of A if $AGA = A$. If A is square and nonsingular then A^{-1} is the unique g-inverse of A. Otherwise, A has infinitely many g-inverses, as we will see shortly.

Let A be an $m \times n$ matrix and let G be a g-inverse of A. If $Ax = b$ is consistent then $x = Gb$ is a solution of $Ax = b$.

Let $A = BC$ be a rank factorization. Then B admits a left inverse B_ℓ^- and C admits a right inverse C_r^-. Then $G = C_r^- B_\ell^-$ is a g-inverse of A, since

$$AGA = BC(C_r^- B_\ell^-)BC = BC = A.$$

Alternatively, if A has rank r then there exist nonsingular matrices P, Q such that

$$A = P\begin{bmatrix} I_r & 0 \\ 0 & 0 \end{bmatrix}Q.$$

It can be verified that for any U, V, W of appropriate dimensions,

$$\begin{bmatrix} I_r & U \\ V & W \end{bmatrix}$$

is a g-inverse of

$$\begin{bmatrix} I_r & 0 \\ 0 & 0 \end{bmatrix}.$$

Then

$$G = Q^{-1}\begin{bmatrix} I_r & U \\ V & W \end{bmatrix}P^{-1}$$

is a g-inverse of A. This also shows that any matrix that is not a square, nonsingular matrix admits infinitely many g-inverses.

Another method that is particularly suitable for computing a g-inverse is as follows. Let A be of rank r. Choose any $r \times r$ nonsingular submatrix of A. For convenience let us assume

$$A = \begin{bmatrix} A_{11} & A_{12} \\ A_{21} & A_{22} \end{bmatrix},$$

where A_{11} is $r \times r$ and nonsingular. Since A has rank r, there exists a matrix X such that $A_{12} = A_{11}X$, $A_{22} = A_{21}X$. Now it can be verified that the $n \times m$ matrix G defined as

$$G = \begin{bmatrix} A_{11}^{-1} & 0 \\ 0 & 0 \end{bmatrix}$$

is a g-inverse of A. (Just multiply AGA out to see this.) We will often use the notation A^- to denote a g-inverse of A.

A g-inverse of A is called a *reflexive g-inverse* if it also satisfies $GAG = G$. Observe that if G is any g-inverse of A then GAG is a reflexive g-inverse of A.

Let A be an $m \times n$ matrix, G be a g-inverse of A and y be in the column space of A. Then the class of solutions of $Ax = y$ is given by $Gy + (I - GA)z$, where z is arbitrary.

A g-inverse G of A is said to be a *minimum norm* g-inverse of A if, in addition to $AGA = A$, it satisfies $(GA)' = GA$. If G is a minimum norm g-inverse of A, then for any y in the column space of A, $x = Gy$ is a solution of $Ax = y$ with minimum norm. A proof of this fact will be given in Chap. 9.

A g-inverse G of A is said to be a *least squares* g-inverse of A if, in addition to $AGA = A$, it satisfies $(AG)' = AG$. If G is a least squares g-inverse of A then for any x, y, $||AGy - y|| \leq ||Ax - y||$.

Moore–Penrose Inverse

If G is a reflexive g-inverse of A that is both minimum norm and least squares then it is called a *Moore–Penrose inverse* of A. In other words, G is a Moore–Penrose inverse of A if it satisfies

$$AGA = A, \quad GAG = G, \quad (AG)' = AG, \quad (GA)' = GA. \tag{1.7}$$

We will show that such a G exists and is, in fact, unique. We first show uniqueness. Suppose G_1, G_2 both satisfy (1.7). Then we must show $G_1 = G_2$. The derivation is as follows.

$$G_1 = G_1AG_1 = G_1G_1'A' = G_1G_1'A'G_2'A' = G_1G_1'A'AG_2$$
$$= G_1AG_1AG_2 = G_1AG_2 = G_1AG_2AG_2 = G_1AA'G_2'G_2$$
$$= A'G_1'A'G_2'G_2 = A'G_2'G_2 = G_2AG_2 = G_2.$$

We will denote the Moore–Penrose inverse of A by A^+. We now show the existence. Let $A = BC$ be a rank factorization. Then it can be easily verified that

$$B^+ = (B'B)^{-1}B', \quad C^+ = C'(CC')^{-1}$$

and then

$$A^+ = C^+B^+.$$

Let A be a symmetric $n \times n$ matrix and let P be an orthogonal matrix such that

$$A = P \, \text{diag}(\lambda_1, \dots, \lambda_n)P'.$$

If $\lambda_1, \dots, \lambda_r$ are the nonzero eigenvalues then

$$A^+ = P \, \text{diag}\left(\frac{1}{\lambda_1}, \dots, \frac{1}{\lambda_r}, 0, \dots, 0\right)P'.$$

In particular, if A is positive semidefinite, then so is A^+.

1.4 Graphs

We assume familiarity with basic theory of graphs. A graph G consists of a finite set of vertices $V(G)$ and a set of edges $E(G)$ consisting of distinct, unordered pairs of vertices. We usually take $V(G)$ to be $\{1, \dots, n\}$ and $E(G)$ to be $\{e_1, \dots, e_m\}$. We may refer to edges $j_1, j_2, \dots$ when we actually mean edges $e_{j_1}, e_{j_2}, \dots$. We consider simple graphs, that is, graphs without loops and parallel edges. Our emphasis is on undirected graphs. However, we do consider directed graphs as well.

If e_k is an edge with end-vertices i and j, then we say that e_k and i or e_k and j are incident. We also write $e_k = \{i, j\}$. The notation $i \sim j$ is used to indicate that i and j are joined by an edge, or that they are adjacent.

Notions such as connected graph, subgraph, degree, path, cycle and so on are standard and will not be recalled here. The complement of the graph G will be denoted by G^c. The complete graph on n vertices will be denoted by K_n. The complete bipartite graph with partite sets of cardinality m, n, will be denoted by $K_{m,n}$. Note that $K_{1,n}$ is called a *star*. Further notions will be recalled as and when the need arises.

Exercises

1. Let A be an $m \times n$ matrix. Show that A and $A'A$ have the same null space. Hence conclude that rank $A = $ rank $A'A$.
2. Let A be a matrix in partitioned form:

$$A = \begin{bmatrix} A_{11} & 0 & \cdots & 0 \\ A_{21} & A_{22} & \cdots & 0 \\ \vdots & & \ddots & \vdots \\ A_{k1} & A_{k2} & \cdots & A_{kk} \end{bmatrix}.$$

   Show that rank $A \geq $ rank $A_{11} + \cdots + $ rank A_{kk}, and that equality holds if $A_{ij} = 0$, $i > j$.
3. Let A be an orthogonal $n \times n$ matrix. Show that a_{11} and det $A(1|1)$ have the same absolute value.
4. Let A and G be matrices of order $m \times n$ and $n \times m$, respectively. Show that $G = A^+$ if and only if $A'AG = A'$ and $G'GA = G'$.
5. If A is a matrix of rank 1, then show that $A^+ = \alpha A'$ for some α. Determine α.

It would be difficult to list the many excellent books that provide the necessary background outlined in this chapter. A few selected references are indicated below.

The books [Bap00] and [HJ85] contain the required matrix theory preliminaries, while [BM08] and [Wes02] are standard introductions to graph theory. The books [BG03] and [CM79] are comprehensive references on generalized inverses.

References and Further Reading

[Bap00] Bapat, R.B.: Linear Algebra and Linear Models, 2nd edn. Hindustan Book Agency, New Delhi, and Springer, Heidelberg (2000)
[BG03]   Ben-Israel, A., Greville, T.N.E.: Generalized Inverses: Theory and Applications, 2nd edn. Springer, New York (2003)
[BM08]   Bondy, J.A., Murty, U.S.R.: Graph Theory, Graduate Texts in Mathematics, 244. Springer, New York (2008)
[CM79]   Campbell, S.L., Meyer, C.D.: Generalized Inverses of Linear Transformation. Pitman, London (1979)
[HJ85]   Horn, R.A., Johnson, C.R.: Matrix Analysis. Cambridge University Press, Cambridge (1985)
[Wes02] West, D.: Introduction to Graph Theory, 2nd edn. Prentice-Hall, New Delhi (2002)

Chapter 2
Incidence Matrix

Let G be a graph with $V(G) = \{1, \ldots, n\}$ and $E(G) = \{e_1, \ldots, e_m\}$. Suppose each edge of G is assigned an orientation, which is arbitrary but fixed. The (*vertex-edge*) *incidence matrix* of G, denoted by $Q(G)$, is the $n \times m$ matrix defined as follows. The rows and the columns of $Q(G)$ are indexed by $V(G)$ and $E(G)$, respectively. The (i, j)-entry of $Q(G)$ is 0 if vertex i and edge e_j are not incident, and otherwise it is 1 or -1 according as e_j originates or terminates at i, respectively. We often denote $Q(G)$ simply by Q. Whenever we mention $Q(G)$ it is assumed that the edges of G are oriented.

Example 2.1 Consider the graph shown. Its incidence matrix is given by Q.

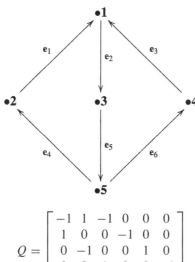

$$Q = \begin{bmatrix} -1 & 1 & -1 & 0 & 0 & 0 \\ 1 & 0 & 0 & -1 & 0 & 0 \\ 0 & -1 & 0 & 0 & 1 & 0 \\ 0 & 0 & 1 & 0 & 0 & -1 \\ 0 & 0 & 0 & 1 & -1 & 1 \end{bmatrix}$$

© Springer-Verlag London 2014
R.B. Bapat, *Graphs and Matrices*, Universitext,
DOI 10.1007/978-1-4471-6569-9_2

2.1 Rank

For any graph G, the column sums of $Q(G)$ are zero and hence the rows of $Q(G)$ are linearly dependent. We now proceed to determine the rank of $Q(G)$.

Lemma 2.2 *If G is a connected graph on n vertices, then* rank $Q(G) = n - 1$.

Proof Suppose x is a vector in the left null space of $Q := Q(G)$, that is, $x'Q = 0$. Then $x_i - x_j = 0$ whenever $i \sim j$. It follows that $x_i = x_j$ whenever there is an ij-path. Since G is connected, x must have all components equal. Thus, the left null space of Q is at most one-dimensional and therefore the rank of Q is at least $n - 1$. Also, as observed earlier, the rows of Q are linearly dependent and therefore rank $Q \leq n - 1$. Hence, rank $Q = n - 1$.   □

Theorem 2.3 *If G is a graph on n vertices and has k connected components then* rank $Q(G) = n - k$.

Proof Let $G_1, \ldots, G_k$ be the connected components of G. Then, after a relabeling of vertices (rows) and edges (columns) if necessary, we have

$$Q(G) = \begin{bmatrix} Q(G_1) & 0 & \cdots & 0 \\ 0 & Q(G_2) & & 0 \\ \vdots & & \ddots & \vdots \\ 0 & 0 & \cdots & Q(G_k) \end{bmatrix}.$$

Since G_i is connected, rank $Q(G_i)$ is $n_i - 1$, where n_i is the number of vertices in $G_i, i = 1, \ldots, k$. It follows that

$$\begin{aligned} \text{rank } Q(G) &= \text{rank } Q(G_1) + \cdots + \text{rank } Q(G_k) \\ &= (n_1 - 1) + \cdots + (n_k - 1) \\ &= n_1 + \cdots + n_k - k = n - k. \end{aligned}$$

This completes the proof.

Lemma 2.4 *Let G be a connected graph on n vertices. Then the column space of $Q(G)$ consists of all vectors $x \in \mathbb{R}^n$ such that $\sum_i x_i = 0$.*

Proof Let U be the column space of $Q(G)$ and let

$$W = \left\{ x \in \mathbb{R}^n : \sum_{i=1}^n x_i = 0 \right\}.$$

Then dim $W = n - 1$. Each column of $Q(G)$ is clearly in W and hence $U \subset W$. It follows by Lemma 2.2 that

$$n - 1 = \dim U \le \dim W = n - 1.$$

Therefore, $\dim U = \dim W$. Thus, $U = W$ and the proof is complete. □

Lemma 2.5 *Let G be a graph on n vertices. Columns $j_1, \ldots, j_k$ of $Q(G)$ are linearly independent if and only if the corresponding edges of G induce an acyclic graph.*

Proof Consider the edges $j_1, \ldots, j_k$ and suppose there is a cycle in the corresponding induced subgraph. Without loss of generality, suppose the columns $j_1, \ldots, j_p$ form a cycle. After relabeling the vertices if necessary, we see that the submatrix of $Q(G)$ formed by the columns $j_1, \ldots, j_p$ is of the form $\begin{bmatrix} B \\ 0 \end{bmatrix}$, where B is the $p \times p$ incidence matrix of the cycle formed by the edges $j_1, \ldots, j_p$. Note that B is a square matrix with column sums zero. Thus, B is singular and the columns $j_1, \ldots, j_p$ are linearly dependent. This proves the "only if" part of the lemma.

Conversely, suppose the edges $j_1, \ldots, j_k$ induce an acyclic graph, that is, a forest. If the forest has q components then clearly $k = n - q$, which by Theorem 2.3, is the rank of the submatrix formed by the columns $j_1, \ldots, j_k$. Therefore, the columns $j_1, \ldots, j_k$ are linearly independent.. □

2.2 Minors

A matrix is said to be *totally unimodular* if the determinant of any square submatrix of the matrix is either 0 or ± 1. It is easily proved by induction on the order of the submatrix that $Q(G)$ is totally unimodular as seen in the next result.

Lemma 2.6 *Let G be a graph with incidence matrix $Q(G)$. Then $Q(G)$ is totally unimodular.*

Proof Consider the statement that any $k \times k$ submatrix of $Q(G)$ has determinant 0 or ± 1. We prove the statement by induction on k. Clearly the statement holds for $k = 1$, since each entry of $Q(G)$ is either 0 or ± 1. Assume the statement to be true for $k - 1$ and consider a $k \times k$ submatrix B of $Q(G)$. If each column of B has a 1 and a -1, then $\det B = 0$. Also, if B has a zero column, then $\det B = 0$. Now suppose B has a column with only one nonzero entry, which must be ± 1. Expand the determinant of B along that column and use induction assumption to conclude that $\det B$ must be 0 or ± 1. □

Lemma 2.7 *Let G be a tree on n vertices. Then any submatrix of $Q(G)$ of order $n - 1$ is nonsingular.*

Proof Consider the submatrix X of $Q(G)$ formed by the rows $1, \ldots, n - 1$. If we add all the rows of X to the last row, then the last row of X becomes the negative of the last row of $Q(G)$. Thus, if Y denotes the submatrix of $Q(G)$ formed by the rows $1, \ldots, n - 2, n$, then $\det X = -\det Y$. Thus, if $\det X = 0$, then $\det Y = 0$.

Continuing this way we can show that if $\det X = 0$ then each $(n-1) \times (n-1)$ submatrix of $Q(G)$ must be singular. In fact, we can show that if any one of the $(n-1) \times (n-1)$ submatrices of $Q(G)$ is singular, then all of them must be so. However, by Lemma 2.2, rank $Q(G) = n-1$ and hence at least one of the $(n-1) \times (n-1)$ submatrices of $Q(G)$ must be nonsingular.   □

We indicate another argument to prove Lemma 2.7. Consider any $n-1$ rows of $Q(G)$. Without loss of generality, we may consider the rows $1, 2, \ldots, n-1$, and let B be the submatrix of $Q(G)$ formed by these rows. Let x be a row vector of $n-1$ components in the row null space of B. Exactly as in the proof of Lemma 2.2, we may conclude that $x_i = 0$ whenever $i \sim n$, and then the connectedness of G shows that x must be the zero vector.

Lemma 2.8 *Let A be an $n \times n$ matrix and suppose A has a zero submatrix of order $p \times q$ where $p + q \geq n + 1$. Then $\det A = 0$.*

Proof Without loss of generality, suppose the submatrix formed by the first p rows and the first q columns of A is the zero matrix. If $p \geq q$, then evaluating $\det A$ by Laplace expansion in terms of the first p rows we see that $\det A = 0$. Similarly, if $p < q$, then by evaluating by Laplace expansion in terms of the first q columns, we see that $\det A = 0$.   □

We return to a general graph G, which is not necessarily a tree. Any submatrix of $Q(G)$ is indexed by a set of vertices and a set of edges. Consider a square submatrix B of $Q(G)$ with the rows corresponding to the vertices $i_1, \ldots, i_k$ and the columns corresponding to the edges $e_{j_1}, \ldots, e_{j_k}$. We call the object formed by these vertices and edges a *substructure* of G. Note that a substructure is not necessarily a subgraph, since one or both end-vertices of some of the edges may not be present in the substructure.

If we take a tree and delete one of its vertices, but not the incident edges, then the resulting substructure will be called a *rootless tree*. In view of Lemma 2.7, the incidence matrix of a rootless tree is nonsingular. Clearly, if we take a vertex-disjoint union of several rootless trees, then the incidence matrix of the resulting substructure is again nonsingular, since it is a direct sum of the incidence matrices of the individual rootless trees.

Example 2.9 The following substructure is a vertex-disjoint union of rootless trees. The deleted vertices are indicated as hollow circles.

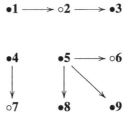

The incidence matrix of the substructure is given by

$$\begin{bmatrix} 1 & 0 & 0 & 0 & 0 & 0 \\ 0 & -1 & 0 & 0 & 0 & 0 \\ 0 & 0 & 1 & 0 & 0 & 0 \\ 0 & 0 & 0 & 1 & 1 & 1 \\ 0 & 0 & 0 & 0 & -1 & 0 \\ 0 & 0 & 0 & 0 & 0 & -1 \end{bmatrix}$$

and is easily seen to be nonsingular. Note that the rows of the incidence matrix are indexed by the vertices $1, 3, 4, 5, 8,$ and 9, respectively.

Let G be a graph with the vertex set $V(G) = \{1, 2, \ldots, n\}$ and the edge set $\{e_1, \ldots, e_m\}$. Consider a submatrix X of $Q(G)$ indexed by the rows $i_1, \ldots, i_k$ and the columns $j_1, \ldots, j_k$. It can be seen that if X is nonsingular then it corresponds to a substructure which is a vertex-disjoint union of rootless trees. A sketch of the argument is as follows. Since X is nonsingular, it does not have a zero row or column. Then, after a relabeling of rows and columns if necessary, we may write

$$X = \begin{bmatrix} X_1 & 0 & \cdots & 0 \\ 0 & X_2 & & 0 \\ \vdots & & \ddots & \\ 0 & 0 & & X_t \end{bmatrix}.$$

If any X_i is not square, then X must have a zero submatrix of order $p \times q$ with $p + q \geq k + 1$. It follows by Lemma 2.8, that $\det X = 0$ and X is singular. Hence, each X_i is a square matrix. Consider the substructure S_i corresponding to X_i. If S_i has a cycle then by Lemma 2.5 X_i is singular. If S_i is acyclic then since, it has an equal number of vertices and edges, it must be a rootless tree.

2.3 Path Matrix

Let G be a graph with the vertex set $V(G) = \{1, 2, \ldots, n\}$ and the edge set $E(G) = \{e_1, \ldots, e_m\}$. Given a path $\mathscr{P}$ in G, the incidence vector of $\mathscr{P}$ is an $m \times 1$ vector defined as follows. The entries of the vector are indexed by $E(G)$. If $e_i \in E(G)$ then the ith element of the vector is 0 if the path does not contain e_i. If the path contains e_i then the entry is 1 or -1, according as the direction of the path agrees or disagrees, respectively, with e_i.

Let G be a tree with the vertex set $\{1, 2, \ldots, n\}$. We identify a vertex, say n, as the root. The path matrix P_n of G (with reference to the root n) is defined as follows. The jth column of P_n is the incidence vector of the (unique) path from vertex j to n, $j = 1, \ldots, n - 1$.

Theorem 2.10 *Let G be a tree with the vertex set $\{1, 2, \ldots, n\}$. Let Q be the incidence matrix of G and let Q_n be the reduced incidence matrix obtained by deleting row n of Q. Then $Q_n^{-1} = P_n$.*

Proof Let $m = n - 1$. For $i \neq j$, consider the (i, j)-element of $P_n Q_n$, which is $\sum_{k=1}^m p_{ik} q_{kj}$. Suppose e_i is directed from x to y, and e_j is directed from w to z. Then $q_{kj} = 0$ unless $k = w$ or $k = z$. Thus,

$$\sum_{k=1}^m p_{ik} q_{kj} = p_{iw} q_{wj} + p_{iz} q_{zj}.$$

As $i \neq j$, we see that the path from w to n contains e_i if and only if the path from z to n contains e_i. Furthermore, when p_{iw} and p_{iz} are nonzero, they both have the same sign. Since $q_{wj} = 1 = -q_{zj}$, it follows that $\sum_{k=1}^m p_{ik} q_{kj} = 0$.

If $i = j$, then we leave it as an exercise to check that $\sum_{k=1}^m p_{ik} q_{ki} = 1$. This completes the proof. $\square$

2.4 Integer Generalized Inverses

An integer matrix need not admit an integer g-inverse. A trivial example is a matrix with each entry equal to 2. Certain sufficient conditions for an integer matrix to have at least one integer generalized inverse are easily given. We describe some such conditions and show that the incidence matrix of a graph belongs to the class.

A square integer matrix is called *unimodular* if its determinant is ± 1.

Lemma 2.11 *Let A be an $n \times n$ integer matrix. Then A is nonsingular and admits an integer inverse if and only if A is unimodular.*

Proof If $\det A = \pm 1$, then $\dfrac{1}{\det A} \operatorname{adj} A$ is the integer inverse of A. Conversely, if A^{-1} exists and is an integer matrix, then from $AA^{-1} = I$ we see that $(\det A)(\det A^{-1}) = 1$ and hence $\det A = \pm 1$. $\square$

The next result gives the well-known Smith normal form of an integer matrix.

Theorem 2.12 *Let A be an $m \times n$ integer matrix. Then there exist unimodular matrices S and T of order $m \times m$ and $n \times n$, respectively, such that*

$$SAT = \begin{bmatrix} \operatorname{diag}(z_1, \ldots, z_r) & 0 \\ 0 & 0 \end{bmatrix},$$

where $z_1, \ldots, z_r$ are positive integers (called the invariant factors of A) such that z_i divides z_{i+1}, $i = 1, 2, \ldots, r - 1$. Furthermore, $z_1 \ldots z_i = d_i$, where d_i is the greatest common divisor of all $i \times i$ minors of A, $i = 1, \ldots, \min\{m, n\}$.

In Theorem 2.12 suppose each $z_i = 1$. Then it is easily verified that $T \begin{bmatrix} I_r & 0 \\ 0 & 0 \end{bmatrix} S$ is an integer g-inverse of A.

Note that if A is an integer matrix which has integer rank factorization $A = FH$, where F admits an integer left inverse F^- and H admits an integer right inverse H^-, then $H^- F^-$ is an integer g-inverse of A.

We denote the column vector consisting of all 1 s by $\mathbf{1}$. The order of the vector will be clear from the context. Similarly the matrix of all 1 s will be denoted by J. We may indicate the $n \times n$ matrix of all 1 s by J_n as well.

In the next result we state the Smith normal form and an integer rank factorization of the incidence matrix explicitly.

Theorem 2.13 *Let G be a graph with vertex set $V(G) = \{1, 2, \ldots, n\}$ and edge set $\{e_1, \ldots, e_m\}$. Suppose the edges $e_1, \ldots, e_{n-1}$ form a spanning tree of G. Let Q_1 be the submatrix of Q formed by the rows $1, \ldots, n-1$ and the columns $e_1, \ldots, e_{n-1}$. Let $q = m - n + 1$. Partition Q as follows:*

$$Q = \begin{bmatrix} Q_1 & Q_1 N \\ -\mathbf{1}'Q_1 & -\mathbf{1}'Q_1 N \end{bmatrix}.$$

Set

$$B = \begin{bmatrix} Q_1^{-1} & 0 \\ 0 & 0 \end{bmatrix},$$

$$S = \begin{bmatrix} Q_1^{-1} & 0 \\ \mathbf{1}' & 1 \end{bmatrix}, \quad T = \begin{bmatrix} I_{n-1} & -N \\ 0 & I_q \end{bmatrix},$$

$$F = \begin{bmatrix} Q_1 \\ -\mathbf{1}'Q_1 \end{bmatrix}, \quad H = \begin{bmatrix} I_{n-1} & N \end{bmatrix}.$$

Then the following assertions hold:

(i) *B is an integer reflexive g-inverse of Q.*

(ii) *S and T are unimodular matrices.*

(iii) *$SQT = \begin{bmatrix} I_{n-1} & 0 \\ 0 & 0 \end{bmatrix}$ is the Smith normal form of Q.*

(iv) *$Q = FH$ is an integer rank factorization of Q.*

The proof of Theorem 2.13 is by a simple verification and is omitted. Also note that F admits an integer left inverse and H admits an integer right inverse.

2.5 Moore–Penrose Inverse

We now turn our attention to the Moore–Penrose inverse Q^+ of Q. We first prove some preliminary results. The next result is the well-known fact that the null space of A^+ is the same as that of A' for any matrix A. We include a proof.

Lemma 2.14 *If A is an $m \times n$ matrix, then for an $n \times 1$ vector x, $Ax = 0$ if and only if $x'A^+ = 0$.*

Proof If $Ax = 0$ then $A^+Ax = 0$ and hence $x'(A^+A)' = 0$. Since A^+A is symmetric, it follows that $x'A^+A = 0$. Hence, $x'A^+AA^+ = 0$, and it follows that $x'A^+ = 0$. The converse follows since $(A^+)^+ = A$.   □

Lemma 2.15 *If G is connected, then $I - QQ^+ = \frac{1}{n}J$.*

Proof Note that $(I - QQ^+)Q = 0$. Thus, any row of $I - QQ^+$ is in the left null space of Q. Since G is connected, the left null space of Q is spanned by the vector $\mathbf{1}'$. Thus, any row of $I - QQ^+$ is a multiple of any other row. Since $I - QQ^+$ is symmetric, it follows that all the elements of $I - QQ^+$ are equal to a constant. The constant must be nonzero, since Q cannot have a right inverse. Now using the fact that $I - QQ^+$ is idempotent, it follows that it must equal $\frac{1}{n}J$.   □

Let G be a graph with $V(G) = \{1, 2, \ldots, n\}$ and $E(G) = \{e_1, \ldots, e_m\}$. Suppose the edges $e_1, \ldots, e_{n-1}$ form a spanning tree of G. Partition Q as follows:

$$Q = \begin{bmatrix} U & V \end{bmatrix},$$

where U is $n \times (n - 1)$ and V is $n \times (m - n + 1)$. Also, let Q^+ be partitioned as

$$Q^+ = \begin{bmatrix} X \\ Y \end{bmatrix},$$

where X is $(n - 1) \times n$ and Y is $(m - n + 1) \times n$.

There exists an $(n - 1) \times (m - n + 1)$ matrix D such that $V = UD$. By Lemma 2.14 it follows that $Y = D'X$. Let $M = I - \frac{1}{n}J$. By Lemma 2.15

$$M = QQ^+ = UX + VY = UX + UDD'X = U(I + DD')X.$$

Thus, for any i, j,
$$U_i(I + DD')X^j = M(i, j),$$

where U_i is U with row i deleted, and X^j is X with column j deleted.

By Lemma 2.7, U_i is nonsingular. Also, DD' is positive semidefinite and thus $I + DD'$ is nonsingular. Therefore, $U_i(I + DD')$ is nonsingular and

$$X^j = (U_i(I + DD'))^{-1}M(i, j).$$

Once X^j is determined, the jth column of X is obtained using the fact that $Q^+\mathbf{1} = 0$. Then Y is determined, since $Y = D'X$.

We illustrate the above method of calculating Q^+ by an example. Consider the graph

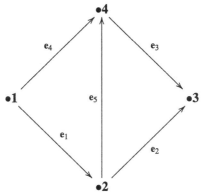

with the incidence matrix

$$
\begin{bmatrix}
1 & 0 & 0 & 1 & 0 \\
-1 & 1 & 0 & 0 & 1 \\
0 & -1 & -1 & 0 & 0 \\
0 & 0 & 1 & -1 & -1
\end{bmatrix}.
$$

Fix the spanning tree formed by $\{e_1, e_2, e_3\}$. Then $Q = \begin{bmatrix} U & V \end{bmatrix}$ where U is formed by the first three columns of Q. Observe that $V = UD$, where

$$
D = \begin{bmatrix}
1 & 0 \\
1 & 1 \\
-1 & -1
\end{bmatrix}.
$$

Set $i = j = 4$. Then $Q^+ = \begin{bmatrix} X \\ Y \end{bmatrix}$ where

$$
X^4 = (U_4(I + DD'))^{-1}M(4, 4) = \frac{1}{8}\begin{bmatrix}
3 & -2 & -1 \\
1 & 2 & -3 \\
1 & 0 & -3
\end{bmatrix}.
$$

The last column of X is found using the fact that the row sums of X are zero. Then $Y = D'X$. After these calculations we see that

$$
Q^+ = \begin{bmatrix} X \\ Y \end{bmatrix} = \frac{1}{8}\begin{bmatrix}
3 & -2 & -1 & 0 \\
1 & 2 & -3 & 0 \\
1 & 0 & -3 & 2 \\
3 & 0 & -1 & -2 \\
0 & 2 & 0 & -2
\end{bmatrix}.
$$

2.6 0–1 Incidence Matrix

We now consider the incidence matrix of an undirected graph. Let G be a graph with $V(G) = \{1, \ldots, n\}$ and $E(G) = \{e_1, \ldots, e_m\}$. The (vertex-edge) incidence matrix of G, which we denote by $M(G)$, or simply by M, is the $n \times m$ matrix defined as follows. The rows and the columns of M are indexed by $V(G)$ and $E(G)$, respectively. The (i, j)-entry of M is 0 if vertex i and edge e_j are not incident, and otherwise it is 1. We often refer to M as the 0–1 incidence matrix for clarity. The proof of the next result is easy and is omitted.

Lemma 2.16 *Let C_n be the cycle on the vertices $\{1, \ldots, n\}$, $n \geq 3$, and let M be its incidence matrix. Then* det *M equals 0 if n is even and 2 if n is odd.*

Lemma 2.17 *Let G be a connected graph with n vertices and let M be the incidence matrix of G. Then the rank of M is $n - 1$ if G is bipartite and n otherwise.*

Proof Suppose $x \in \mathbb{R}^n$ such that $x'M = 0$. Then $x_i + x_j = 0$ whenever the vertices i and j are adjacent. Since G is connected it follows that $|x_i| = \alpha$, $i = 1, \ldots, n$, for some constant α. Suppose G has an odd cycle formed by the vertices $i_1, \ldots, i_k$. Then going around the cycle and using the preceding observations we find that $\alpha = -\alpha$ and hence $\alpha = 0$. Thus, if G has an odd cycle then the rank of M is n.

Now suppose G has no odd cycle, that is, G is bipartite. Let $V(G) = X \cup Y$ be a bipartition. Orient each edge of G giving it the direction from X to Y and let Q be the corresponding $\{0, 1, -1\}$-incidence matrix. Note that Q is obtained from M by multiplying the rows corresponding to the vertices in Y by -1. Consider the columns $j_1, \ldots, j_{n-1}$ corresponding to a spanning tree of G and let B be the submatrix formed by these columns. By Lemma 2.7 any $n - 1$ rows of B are linearly independent and (since rows of M and Q coincide up to a sign) the corresponding rows of M are also linearly independent. Thus, rank $M \geq n - 1$.

Let $z \in \mathbb{R}^n$ be the vector with z_i equal to 1 or -1 according as i belongs to X or to Y, respectively. Then it is easily verified that $z'M = 0$ and thus the rows of M are linearly dependent. Thus, rank $M = n - 1$ and the proof is complete. $\square$

A connected graph is said to be *unicyclic* if it contains exactly one cycle. We omit the proof of the next result, since it is based on arguments as in the oriented case.

Lemma 2.18 *Let G be a graph and let R be a substructure of G with an equal number of vertices and edges. Let N be the incidence matrix of R. Then N is nonsingular if and only if R is a vertex-disjoint union of rootless trees and unicyclic graphs with the cycle being odd.*

We summarize some basic properties of the minors of the incidence matrix of an undirected graph.

Let M be the 0–1 incidence matrix of the graph G with n vertices. Let N be a square submatrix of M indexed by the vertices and edges, which constitute a substructure denoted by R. If N has a zero row or a zero column then, clearly, det $N = 0$. This case corresponds to R having an isolated vertex or an edge with both endpoints missing. We assume this not to be the case.

Let R be the vertex-disjoint union of the substructures $R_1, \ldots, R_k$. After a relabeling of rows and columns if necessary, we have

$$
N = \begin{bmatrix} N_1 & 0 & \cdots & 0 \\ 0 & N_2 & & 0 \\ \vdots & & \ddots & \\ 0 & 0 & & N_k \end{bmatrix},
$$

where N_i is the incidence matrix of R_i, $i = 1, \ldots, k$.

If N_i is not square for some i, then using Lemma 2.8, we conclude that N is singular. Thus, if R_i has unequal number of vertices and edges for some i then $\det N = 0$.

If R_i is unicyclic for some i, with the cycle being even, then $\det N = 0$. This follows easily from Lemma 2.16.

Now suppose each N_i is square. Then each R_i is either a rootless tree or is unicyclic with the cycle being odd. In the first case, $\det N_i = \pm 1$ while in the second case $\det N_i = \pm 2$. Note that $\det N = \prod_{i=1}^{k} \det N_i$, Thus, in this case $\det N = \pm 2^{\omega_1(R)}$, where $\omega_1(R)$ is the number of substructures $R_1, \ldots, R_k$ that are unicyclic.

The concept of a substructure will not be needed extensively henceforth. It seems essential to use the concept if one wants to investigate minors of incidence matrices. We have not developed the idea rigorously and have tried to use it informally.

2.7 Matchings in Bipartite Graphs

Lemma 2.19 *Let G be a bipartite graph. Then the 0–1 incidence matrix M of G is totally unimodular.*

Proof The proof is similar to that of Lemma 2.6. Consider the statement that any $k \times k$ submatrix of M has determinant 0 or ± 1. We prove the statement by induction on k. Clearly the statement holds for $k = 1$, since each entry of M is either 0 or 1. Assume the statement to be true for $k - 1$ and consider a $k \times k$ submatrix B of M. If B has a zero column, then $\det B = 0$. Suppose B has a column with only one nonzero entry, which must be 1. Expand the determinant of B along that column and use the induction assumption to conclude that $\det B$ must be 0 or ± 1. Finally, suppose each column of B has two nonzero entries. Let $V(G) = X \cup Y$ be the bipartition of G. The sum of the rows of B corresponding to the vertices in X must equal the sum of the rows of B corresponding to the vertices in Y. In fact both these sums will be $\mathbf{1}'$. Therefore, B is singular in this case and $\det B = 0$. This completes the proof. □

Recall that a matching in a graph is a set of edges, no two of which have a vertex in common. The matching number $\nu(G)$ of the graph G is defined to be the maximum number of edges in a matching of G.

We need some background from the theory of linear inequalities and linear programming in the following discussion.

Let G be a graph with $V(G) = \{1, \ldots, n\}$, $E(G) = \{e_1, \ldots, e_m\}$. Let M be the incidence matrix of G. Note that a 0–1 vector x of order $m \times 1$ is the incidence vector of a matching if and only if it satisfies $Mx \leq \mathbf{1}$. Consider the linear programming problem:

$$\max\{\mathbf{1}'x\} \text{ subject to } x \geq 0, \quad Mx \leq \mathbf{1}. \tag{2.1}$$

In order to solve (2.1) we may restrict attention to the basic feasible solutions, which are constructed as follows. Let rank $M = r$. Find a nonsingular $r \times r$ submatrix B of M and let $y = B^{-1}\mathbf{1}$. Set the subvector of x corresponding to the rows in B equal to y and set the remaining coordinates of x equal to 0. If the x thus obtained satisfies $x \geq 0$, $Mx \leq \mathbf{1}$, then it is called a basic feasible solution. With this terminology and notation we have the following.

Lemma 2.20 *Let G be a bipartite graph with incidence matrix M. Then there exists a 0–1 vector z which is a solution of* (2.1).

Proof By Lemma 2.19, M is totally unimodular and hence for any nonsingular submatrix B of M, B^{-1} is an integral matrix. By the preceding discussion, a basic feasible solution of $x \geq 0$, $Mx \leq \mathbf{1}$ has only integral coordinates. Hence there is a nonnegative, integral vector z which solves (2.1). Clearly if a coordinate of z is >1, then z cannot satisfy $Mz \leq \mathbf{1}$. Hence z must be a 0–1 vector. $\square$

A vertex cover in a graph is a set of vertices such that each edge in the graph is incident to one of the vertices in the set. The covering number $\tau(G)$ of the graph G is defined to be the minimum number of vertices in a vertex cover of G.

As before, let G be a graph with $V(G) = \{1, \ldots, n\}$, $E(G) = \{e_1, \ldots, e_m\}$. Let M be the incidence matrix of G. Note that a 0–1 vector x of order $n \times 1$ is the incidence vector of a vertex cover if and only if it satisfies $M'x \geq \mathbf{1}$. Consider the linear programming problem:

$$\min\{\mathbf{1}'x\} \text{ subject to } x \geq 0, \quad M'x \geq \mathbf{1} \tag{2.2}$$

The proof of the next result is similar to that of Lemma 2.20 and hence is omitted.

Lemma 2.21 *Let G be a bipartite graph with the incidence matrix M. Then there exists a 0–1 vector z which is a solution of* (2.2).

The following result is the well-known König–Egervary theorem, which is central to the matching theory of bipartite graphs.

Theorem 2.22 *If G is a bipartite graph then $\nu(G) = \tau(G)$.*

Proof Let M be the incidence matrix of G. The linear programming problems (2.1) and (2.2) are dual to each other and their feasibility is obvious. Hence, by the duality theorem, their optimal values are equal. As discussed earlier, the optimal values of the two problems are $v(G)$ and $\tau(G)$, respectively. Hence it follows that $v(G) = \tau(G)$.

□

Exercises

1. Let G be an oriented graph with the incidence matrix Q, and let B be a $k \times k$ submatrix of Q which is nonsingular. Show that there is precisely one permutation σ of $1, \ldots, k$ for which the product $b_{1\sigma(1)} \ldots b_{k\sigma(k)}$ is nonzero. (The property holds for the 0–1 incidence matrix as well.)
2. Let G be a connected graph with $V(G) = \{1, \ldots, n\}$ and $E(G) = \{e_1, \ldots, e_m\}$. Suppose the edges of G are oriented, and let Q be the incidence matrix. Let y be an $n \times 1$ vector with one coordinate 1, one coordinate -1, and the remaining coordinates zero. Show that there exists an $m \times 1$ vector x with coordinates $0, \pm 1$ such that $Qx = y$. Give a graph-theoretic interpretation.
3. Let each edge of K_n be given an orientation and let Q be the incidence matrix. Determine Q^+.
4. Let M be the 0–1 incidence matrix of the graph G. Show that if M is totally unimodular then G is bipartite.
5. Let A be an $n \times n$ 0–1 matrix. Show that the following conditions are equivalent:

   (i) For any permutation σ of $1, \ldots, n$, $a_{1\sigma(1)} \ldots a_{n\sigma(n)} = 0$.
   (ii) A has a zero submatrix of order $r \times s$ where $r + s = n + 1$.

Biggs [Big93] and Godsil and Royle [GR01] are essential references for the material related to this chapter as well as that in Chaps 3–6. Relevant references for various sections are as follows: Sect. 2.3: [Bap02], Sect. 2.4: [BHK81], Sect. 2.5: [Ij65], Sect. 2.6: [GKS95], Sect. 2.7: [LP86].

References and Further Reading

[Bap02]   Bapat, R.B., Pati, S.: Path matrices of a tree. J. Math. Sci. **1**, 46–52 (2002). New Series (Delhi)
[BHK81]   Bevis, J.H., Hall, F.J., Katz, I.J.: Integer generalized inverses of incidence matrices. Linear Algebra Appl. **39**, 247–258 (1981)
[Big93]   Biggs, N.: Algebraic Graph Theory, 2nd edn. Cambridge University Press, Cambridge (1993)
[GR01]   Godsil, C., Royle, G.: Algebraic Graph Theory, Graduate Texts in Mathematics. Springer, New York (2001)
[GKS95]   Grossman, J.W., Kulkarni, D., Schochetman, I.E.: On the minors of an incidence matrix and its Smith normal form. Linear Algebra Appl. **218**, 213–224 (1995)

[Ij65]   Ijiri, Y.: On the generalized inverse of an incidence matrix. J. Soc. Ind. Appl. Math. **13**(3), 827–836 (1965)
[LP86]   Lovász, L., Plummer, M.D.: Matching Theory, Annals of Discrete Mathematics. North-Holland, Amsterdam (1986)

Chapter 3
Adjacency Matrix

Let G be a graph with $V(G) = \{1, \ldots, n\}$ and $E(G) = \{e_1, \ldots, e_m\}$. The adjacency matrix of G, denoted by $A(G)$, is the $n \times n$ matrix defined as follows. The rows and the columns of $A(G)$ are indexed by $V(G)$. If $i \neq j$ then the (i, j)-entry of $A(G)$ is 0 for vertices i and j nonadjacent, and the (i, j)-entry is 1 for i and j adjacent. The (i, i)-entry of $A(G)$ is 0 for $i = 1, \ldots, n$. We often denote $A(G)$ simply by A.

Example 3.1 Consider the graph G:

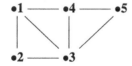

Then

$$A(G) = \begin{bmatrix} 0 & 1 & 1 & 1 & 0 \\ 1 & 0 & 1 & 0 & 0 \\ 1 & 1 & 0 & 1 & 1 \\ 1 & 0 & 1 & 0 & 1 \\ 0 & 0 & 1 & 1 & 0 \end{bmatrix}.$$

Clearly A is a symmetric matrix with zeros on the diagonal. For $i \neq j$, the principal submatrix of A formed by the rows and the columns i, j is the zero matrix if $i \not\sim j$ and otherwise it equals $\begin{bmatrix} 0 & 1 \\ 1 & 0 \end{bmatrix}$. The determinant of this matrix is -1. Thus, the sum of the 2×2 principal minors of A equals $-|E(G)|$.

Consider the principal submatrix of A formed by the three distinct rows and columns, i, j, k. It can be seen that the submatrix is nonsingular only when i, j, k are

© Springer-Verlag London 2014

R.B. Bapat, *Graphs and Matrices*, Universitext,

DOI 10.1007/978-1-4471-6569-9_3

adjacent to each other (i.e., they constitute a triangle). In that case the submatrix is

$$\begin{bmatrix} 0 & 1 & 1 \\ 1 & 0 & 1 \\ 1 & 1 & 0 \end{bmatrix}.$$

The determinant of this matrix is 2. Thus, the sum of the 3×3 principal minors of A equals twice the number of triangles in G.

We make an elementary observation about the powers of A. The (i, j)-entry of A^k is the number of walks of length k from i to j. This is clear from the definition of matrix multiplication.

Let G be a connected graph with vertices $\{1, \ldots, n\}$. The distance $d(i, j)$ between the vertices i and j is defined as the minimum length of an (ij)-path. We set $d(i, i) = 0$. The maximum value of $d(i, j)$ is the diameter of G.

Lemma 3.2 *Let G be a connected graph with vertices $\{1, \ldots, n\}$ and let A be the adjacency matrix of G. If i, j are vertices of G with $d(i, j) = m$, then the matrices $I, A, \ldots, A^m$ are linearly independent.*

Proof We may assume $i \neq j$. There is no (ij)-path of length less than m. Thus, the (i, j)-element of $I, A, \ldots, A^{m-1}$ is zero, whereas the (i, j)-element of A^m is nonzero. Hence, the result follows. $\square$

Corollary 3.3 *Let G be a connected graph with k distinct eigenvalues and let d be the diameter of G. Then $k > d$.*

Proof Let A be the adjacency matrix of G. By Lemma 3.2, the matrices $I, A, \ldots, A^d$ are linearly independent. Thus, the degree of the minimal polynomial of A, which equals k, must exceed d. $\square$

3.1 Eigenvalues of Some Graphs

Let G be a graph with adjacency matrix A. Often we refer to the eigenvalues of A as the eigenvalues of G. We now determine the eigenvalues of some graphs.

Theorem 3.4 *(i) For any positive integer n, the eigenvalues of K_n are $n - 1$ with multiplicity 1 and -1 with multiplicity $n - 1$. (ii) For any positive integers p, q, the eigenvalues of $K_{p,q}$ are $\sqrt{pq}, -\sqrt{pq}$ and 0 with multiplicity $p + q - 2$.*

Proof First consider J_n, the $n \times n$ matrix of all ones. It is a symmetric, rank 1 matrix, and hence it has only one nonzero eigenvalue, which must equal the trace. Thus, the eigenvalues of J_n are n with multiplicity 1 and 0 with multiplicity $n - 1$. Since $A(K_n) = J_n - I_n$, the eigenvalues of $A(K_n)$ must be as asserted in (i).

To prove (ii), note that

$$A(K_{p,q}) = \begin{bmatrix} 0 & J_{pq} \\ J_{qp} & 0 \end{bmatrix},$$

where J_{pq} and J_{qp} are matrices of all ones of the appropriate size. Now

$$\text{rank } A(K_{p,q}) = \text{rank } J_{pq} + \text{rank } J_{qp} = 2,$$

and hence $A(K_{p,q})$ must have precisely two nonzero eigenvalues. These must be of the form λ and $-\lambda$, since the trace of $A(K_{p,q})$ is zero. As noted earlier, the sum of the 2×2 principal minors of $A(K_{p,q})$ is negative the number of edges, that is, $-pq$. This sum also equals the sum of the products of the eigenvalues, taken two at a time, which is $-\lambda^2$. Thus, $\lambda^2 = pq$ and the eigenvalues must be as asserted in (ii). $\qquad\square$

For a positive integer $n \geq 2$, let Q_n be the full-cycle permutation matrix of order n. Thus, the $(i, i+1)$-element of Q_n is $1, i = 1, 2, \ldots, n-1$, the $(n, 1)$-element of Q_n is 1, and the remaining elements of Q_n are zero.

Lemma 3.5 *For $n \geq 2$, the eigenvalues of Q_n are $1, \omega, \omega^2, \ldots, \omega^{n-1}$, where $\omega = e^{\frac{2\pi i}{n}}$, is the primitive nth root of unity.*

Proof The characteristic polynomial of Q_n is $\det(Q_n - \lambda I) = (-1)^n(\lambda^n - 1)$. Clearly, the roots of the characteristic polynomial are the n roots of unity. $\qquad\square$

For a positive integer n, C_n and P_n will denote the cycle and the path on n vertices, respectively.

Theorem 3.6 *For $n \geq 2$, the eigenvalues of C_n are $2\cos\frac{2\pi k}{n}, k = 1, \ldots, n$.*

Proof Note that $A(C_n) = Q_n + Q_n' = Q_n + Q_n^{n-1}$ is a polynomial in Q_n. Thus, the eigenvalues of $A(C_n)$ are obtained by evaluating the same polynomial at each of the eigenvalues of Q_n. Thus, by Lemma 3.5, the eigenvalues of $A(C_n)$ are $\omega^k + \omega^{n-k}$, $k = 1, \ldots, n$. Note that

$$\omega^k + \omega^{n-k} = \omega^k + \omega^{-k}$$
$$= e^{\frac{2\pi ik}{n}} + e^{-\frac{2\pi ik}{n}}$$
$$= 2\cos\frac{2\pi k}{n},$$

$k = 1, \ldots, n$, and the proof is complete. $\qquad\square$

Theorem 3.7 *For $n \geq 1$, the eigenvalues of P_n are $2\cos\frac{\pi k}{n+1}, k = 1, \ldots, n$.*

Proof Let λ be an eigenvalue of $A(P_n)$ with x as the corresponding eigenvector. By symmetry, $(-x_n, -x_{n-1}, \ldots, -x_1)$ is also an eigenvector of $A(P_n)$ for λ.

It may be verified that

$$(x_1, \ldots, x_n, 0, -x_n, \ldots, -x_1, 0)$$

and

$$(0, x_1, \ldots, x_n, 0, -x_n, \ldots, -x_1)$$

are two linearly independent eigenvectors of $A(C_{2n+2})$ for the same eigenvalue. We illustrate this by an example. Suppose $x = (x_1, x_2, x_3)'$ is an eigenvector of $A(P_3)$ for the eigenvalue λ. Then

$$\begin{bmatrix} 0 & 1 & 0 \\ 1 & 0 & 1 \\ 0 & 1 & 0 \end{bmatrix} \begin{bmatrix} x_1 \\ x_2 \\ x_3 \end{bmatrix} = \lambda \begin{bmatrix} x_1 \\ x_2 \\ x_3 \end{bmatrix}.$$

We obtain an eigenvector of $A(C_8)$ for the same eigenvalue, since it may be verified that

$$\begin{bmatrix} 0 & 1 & 0 & 0 & 0 & 0 & 0 & 1 \\ 1 & 0 & 1 & 0 & 0 & 0 & 0 & 0 \\ 0 & 1 & 0 & 1 & 0 & 0 & 0 & 0 \\ 0 & 0 & 1 & 0 & 1 & 0 & 0 & 0 \\ 0 & 0 & 0 & 1 & 0 & 1 & 0 & 0 \\ 0 & 0 & 0 & 0 & 1 & 0 & 1 & 0 \\ 0 & 0 & 0 & 0 & 0 & 1 & 0 & 1 \\ 1 & 0 & 0 & 0 & 0 & 0 & 1 & 0 \end{bmatrix} \begin{bmatrix} x_1 \\ x_2 \\ x_3 \\ 0 \\ -x_3 \\ -x_2 \\ -x_1 \\ 0 \end{bmatrix} = \lambda \begin{bmatrix} x_1 \\ x_2 \\ x_3 \\ 0 \\ -x_3 \\ -x_2 \\ -x_1 \\ 0 \end{bmatrix}.$$

Continuing with the proof, we have established that each eigenvalue of P_n must be an eigenvalue of C_{2n+2} of multiplicity 2. By Theorem 3.6, the eigenvalues of C_{2n+2} are $2\cos\frac{2\pi k}{2n+2} = 2\cos\frac{\pi k}{n+1}$, $k = 1, \ldots, 2n+2$. Of these, the eigenvalues that appear twice, in view of the periodicity of the cosine, are $2\cos\frac{\pi k}{n+1}$, $k = 1, \ldots, n$, which must be the eigenvalues of P_n.   □

3.2 Determinant

We now introduce some definitions. Let G be a graph with $V(G) = \{1, \ldots, n\}$ and edge set $E(G)$. A subgraph H of G is called an elementary subgraph if each component of H is either an edge or a cycle. Denote by $c(H)$ and $c_1(H)$ the number of components in a subgraph H which are cycles and edges, respectively.

Theorem 3.8 *Let G be a graph with $V(G) = \{1, \ldots, n\}$ and let A be the adjacency matrix of G. Then*

$$\det A = \sum (-1)^{n-c_1(H)-c(H)} 2^{c(H)},$$

where the summation is over all spanning elementary subgraphs H of G.

Proof We have

$$\det A = \sum_{\pi} \operatorname{sgn}(\pi) a_{1\pi(1)} \cdots a_{n\pi(n)},$$

where the summation is over all permutations of $1, \ldots, n$. Consider a term

$$a_{1\pi(1)} \cdots a_{n\pi(n)},$$

which is nonzero. Since π admits a cycle decomposition, such a term will correspond to some 2-cycles (ij) of π, which designate an edge joining i and j in G, as well as some cycles of higher order, which correspond to cycles in G. (Note that $\pi(i) \neq i$ for any i.) Thus, each nonzero term in the summation arises from an elementary subgraph of G with vertex set $V(G)$. Suppose the term $a_{1\pi(1)} \cdots a_{n\pi(n)}$ corresponds to the spanning elementary subgraph H. The sign of π is (-1) raised to n minus the number of cycles in the cycle decomposition of π, which is the same as $(-1)^{n-c_1(H)-c(H)}$.

Finally, each spanning elementary subgraph gives rise to $2^{c(H)}$ terms in the summation, since each cycle can be associated to a cyclic permutation in two ways. In view of these observations the proof is complete.   □

Example 3.9 Consider the graph G:

There are three spanning elementary subgraphs of G, given by H_1, H_2 and H_3, where

$$V(H_1) = V(H_2) = V(H_3) = \{1, 2, 3, 4\},$$

and

$$E(H_1) = \{12, 34\}, \quad E(H_2) = \{14, 23\}, \quad E(H_3) = \{12, 23, 34, 41\}.$$

By Theorem 3.8, $\det A = 2(-1)^{4-2-0} 2^0 + (-1)^{4-0-1} 2^1 = 0$. This fact is also evident since A has two identical columns.

Theorem 3.10 *Let G be a graph with vertices $\{1, \ldots, n\}$ and let A be the adjacency matrix of G. Let*

$$\phi_\lambda(A) = \det(\lambda I - A) = \lambda^n + c_1\lambda^{n-1} + \cdots + c_n$$

be the characteristic polynomial of A. Then $c_k = \sum(-1)^{c_1(H)+c(H)}2^{c(H)}$, *where the summation is over all the elementary subgraphs H of G with k vertices, $k = 1, \ldots, n$.*

Proof Observe that c_k is $(-1)^k$ times the sum of the principal minors of A of order k, $k = 1, \ldots, n$. By Theorem 3.8,

$$c_k = (-1)^k \sum(-1)^{k-c_1(H)-c(H)}2^{c(H)},$$

where the summation is over all the elementary subgraphs H of G with k vertices. Hence, c_k is as asserted in the theorem. Note that $c_1 = 0$. □

At the beginning of this chapter we gave an interpretation of c_2 and c_3, which can be regarded as special cases of Theorem 3.10.

Corollary 3.11 *Let G be a graph with vertices $\{1, \ldots, n\}$ and let A be the adjacency matrix of G. Let*

$$\phi_\lambda(A) = \det(\lambda I - A) = \lambda^n + c_1\lambda^{n-1} + \cdots + c_n$$

be the characteristic polynomial of A. Suppose $c_3 = c_5 = \cdots = c_{2k-1} = 0$. Then G has no odd cycle of length i, $3 \le i \le 2k - 1$. Furthermore, the number of $(2k + 1)$-cycles in G is $-\frac{1}{2}c_{2k+1}$.

Proof Since $c_3 = 0$, there are no triangles in G. Thus, any elementary subgraph of G with 5 vertices must only comprise of a 5-cycle. It follows by Theorem 3.10 that if $c_5 = 0$ then there are no 5-cycles in G. Continuing this way we find that if $c_3 = c_5 = \cdots = c_{2k-1} = 0$, then any elementary subgraph of G with $2k+1$ vertices must be a $(2k + 1)$-cycle. Furthermore, by Theorem 3.10,

$$c_{2k+1} = \sum(-1)^{c_1(H)+c(H)}2^{c(H)},$$

where the summation is over all $(2k + 1)$-cycles H in G. For any $(2k + 1)$-cycle H, $c_1(H) = 0$ and $c(H) = 1$. Therefore, c_{2k+1} is (-2) the number of $(2k + 1)$-cycles in G. That completes the proof. □

Corollary 3.12 *Using the notation of Corollary 3.11, if $c_{2k+1} = 0$, $k = 0, 1, \ldots$, then G is bipartite.*

Proof If $c_{2k+1} = 0$, $k = 0, 1, 2, \ldots$, then by Corollary 3.11, G has no odd cycles and hence G must be bipartite. □

We now proceed to show that bipartite graphs can be characterized in terms of the eigenvalues of the adjacency matrix. We first prove the following.

Lemma 3.13 *Let G be a bipartite graph with adjacency matrix A. If λ is an eigenvalue of A with multiplicity k, then $-\lambda$ is also an eigenvalue of A with multiplicity k.*

Proof Let $V(G) = X \cup Y$ be a bipartition of G. We may assume $|X| = |Y|$ by adding isolated vertices if necessary. This does not affect the property we wish to prove, since A only gets changed in the sense that some zero rows and columns are appended. So suppose $|X| = |Y| = m$; then by a relabeling of vertices if necessary, we may write $A = \begin{bmatrix} 0 & B \\ B' & 0 \end{bmatrix}$, where B is $m \times m$. Let x be an eigenvector of A corresponding to λ. Partition x conformally so that we get the equation

$$\begin{bmatrix} 0 & B \\ B' & 0 \end{bmatrix} \begin{bmatrix} x^{(1)} \\ x^{(2)} \end{bmatrix} = \lambda \begin{bmatrix} x^{(1)} \\ x^{(2)} \end{bmatrix}.$$

Then it may be verified that

$$\begin{bmatrix} 0 & B \\ B' & 0 \end{bmatrix} \begin{bmatrix} x^{(1)} \\ -x^{(2)} \end{bmatrix} = -\lambda \begin{bmatrix} x^{(1)} \\ -x^{(2)} \end{bmatrix}.$$

Thus, $-\lambda$ is also an eigenvalue of A. It is also clear that if we have k linearly independent eigenvectors for λ, then the above construction will produce k linearly independent eigenvectors for $-\lambda$. Thus, the multiplicity of $-\lambda$ is also k. That completes the proof. □

Theorem 3.14 *Let G be a graph with vertices $\{1, \ldots, n\}$ and let A be the adjacency matrix of G. Then the following conditions are equivalent.*

(i) *G is bipartite;*
(ii) *if $\phi_\lambda(A) = \lambda^n + c_1 \lambda^{n-1} + \cdots + c_n$ is the characteristic polynomial of A, then $c_{2k+1} = 0, k = 0, 1, \ldots$;*
(iii) *the eigenvalues of A are symmetric with respect to the origin, i.e., if λ is an eigenvalue of A with multiplicity k, then $-\lambda$ is also an eigenvalue of A with multiplicity k.*

Proof The fact that (i) $\Longrightarrow$ (iii) has been proved in Lemma 3.13.

We now show that (iii) $\Longrightarrow$ (ii). Let $\lambda_1, \ldots, \lambda_k, -\lambda_1, \ldots, -\lambda_k$ be the nonzero eigenvalues of A. Here $\lambda_1, \ldots, \lambda_k$ are not necessarily distinct. Then 0 is an eigenvalue of A with multiplicity $n - 2k$. The characteristic polynomial of A equals $\lambda^{n-2k}(\lambda^2 - \lambda_1^2) \cdots (\lambda^2 - \lambda_k^2)$. It follows that $c_{2k+1} = 0, k = 0, 1, \ldots$, and hence (ii) holds.

Finally, it follows from Corollary 3.12, that (ii) $\Longrightarrow$ (i), and the proof is complete. □

3.3 Bounds

We begin with an easy bound for the largest eigenvalue of a graph.

Theorem 3.15 *Let G be a graph with n vertices, m edges and let $\lambda_1 \geq \cdots \geq \lambda_n$ be the eigenvalues of G. Then $\lambda_1 \leq (\frac{2m(n-1)}{n})^{\frac{1}{2}}$.*

Proof As noted earlier, we have $\sum_{i=1}^{n} \lambda_i = 0$ and $\sum_{i=1}^{n} \lambda_i^2 = 2m$. Therefore, $\lambda_1 = -\sum_{i=2}^{n} \lambda_i$ and hence

$$\lambda_1 \leq \sum_{i=2}^{n} |\lambda_i|. \tag{3.1}$$

By the Cauchy–Schwarz inequality and (3.1),

$$2m - \lambda_1^2 = \sum_{i=2}^{n} \lambda_i^2 \geq \frac{1}{n-1} \left(\sum_{i=2}^{n} |\lambda_i| \right)^2 \geq \frac{\lambda_1^2}{n-1}.$$

Hence,

$$2m \geq \lambda_1^2 \left(1 + \frac{1}{n-1} \right) = \lambda_1^2 \left(\frac{n}{n-1} \right)$$

and therefore $\lambda_1^2 \leq \frac{2m(n-1)}{n}$. $\qquad\qquad\square$

We now obtain bounds for the largest and the smallest eigenvalues of a graph in terms of vertex degrees and the chromatic number. Our main tool will be the extremal representation for the largest and the smallest eigenvalues of a symmetric matrix.

Let G be a graph with n vertices and with eigenvalues $\lambda_1 \geq \cdots \geq \lambda_n$. We denote λ_1 and λ_n by $\lambda_1(G)$ and $\lambda_n(G)$, respectively. Similarly, $\lambda_1(B)$ and $\lambda_n(B)$ will denote the largest and the smallest eigenvalues of the symmetric matrix B.

Lemma 3.16 *Let G be a graph with n vertices and let H be an induced subgraph of G with p vertices. Then $\lambda_1(G) \geq \lambda_1(H)$ and $\lambda_n(G) \leq \lambda_p(H)$.*

Proof Note that $A(H)$ is a principal submatrix of $A(G)$. The result follows from the interlacing inequalities relating the eigenvalues of a symmetric matrix and of its principal submatrix. $\qquad\qquad\square$

For a graph G, we denote by $\Delta(G)$ and $\delta(G)$, the maximum and the minimum of the vertex degrees of G, respectively.

Lemma 3.17 *For a graph G, $\delta(G) \leq \lambda_1(G) \leq \Delta(G)$.*

Proof Let A be the adjacency matrix of G and let x be an eigenvector of A corresponding to $\lambda_1(G)$. Then $Ax = \lambda_1(G)x$. From the ith equation of this vector equation we get

$$\lambda_1(G)x_i = \sum_{j \sim i} x_j, i = 1, \ldots, n. \tag{3.2}$$

Let $x_k > 0$ be the maximum coordinate of x. Then from (3.2),

$$\lambda_1(G)x_k = \sum_{j \sim k} x_j \leq \Delta(G)x_k,$$

and hence $\lambda_1(G) \leq \Delta(G)$.

To prove the lower bound, first recall the extremal representation

$$\lambda_1(A) = \max_{||x||=1} \{x'Ax\} = \max_{x \neq 0} \left\{ \frac{x'Ax}{x'x} \right\}.$$

It follows by the extremal representation that

$$\lambda_1(G) \geq \frac{1'A1}{1'1} = \frac{2m}{n}, \tag{3.3}$$

where m is the number of edges in G.

If $d_1, \ldots, d_n$ are the vertex degrees of G, then $2m = d_1 + \cdots + d_n \geq n\delta(G)$ and it follows from (3.3) that $\lambda_1(G) \geq \delta(G)$. $\qquad\square$

Recall that the chromatic number $\chi(G)$ of a graph G is the minimum number of colours required to colour the vertices so that adjacent vertices get distinct colours (such a colouring is called a *proper colouring*). The following result is attributed to Wilf.

Theorem 3.18 *For any graph G, $\chi(G) \leq 1 + \lambda_1(G)$.*

Proof The result is trivial if $\chi(G) = 1$. Let $\chi(G) = p \geq 2$. Let H be an induced subgraph of G such that $\chi(H) = p$ and furthermore, suppose H is minimal with respect to the number of vertices. That is to say, $\chi(H \setminus \{i\}) < p$ for any vertex i of H.

We claim that $\delta(H) \geq p - 1$. Indeed, suppose i is a vertex of H with degree less than $p - 1$. Since $\chi(H \setminus \{i\}) < p$, we may properly colour vertices of $H \setminus \{i\}$ with $p - 1$ colours. Since the degree of i is less than $p - 1$, we may extend the colouring to a proper $(p - 1)$-colouring of H, a contradiction. Hence the degree of each vertex of H is at least $p - 1$ and therefore $\delta(H) \geq p - 1$.

By Lemmas 3.16 and 3.17 we have

$$\lambda_1(G) \geq \lambda_1(H) \geq \delta(H) \geq p - 1$$

and hence $\lambda_1(G) \geq p - 1$. $\qquad\square$

We now prove some results in preparation of the next bound involving chromatic number and eigenvalues.

Lemma 3.19 *If B and C are symmetric $n \times n$ matrices, then*

$$\lambda_1(B + C) \leq \lambda_1(B) + \lambda_1(C).$$

Proof By the extremal representation of the maximum eigenvalue of a symmetric matrix,

$$
\begin{aligned}
\lambda_1(B + C) &= \max_{||x||=1} \{x'(B + C)x\} \\
&\leq \max_{||x||=1} \{x'Bx\} + \max_{||x||=1} \{x'Cx\} \\
&= \lambda_1(B) + \lambda_1(C).
\end{aligned}
$$

This completes the proof.   □

Lemma 3.20 *Let B be an $n \times n$ positive semidefinite matrix and suppose B is partitioned as*

$$
B = \begin{bmatrix} B_{11} & B_{12} \\ B_{21} & B_{22} \end{bmatrix},
$$

where B_{11} is $p \times p$. Then $\lambda_1(B) \leq \lambda_1(B_{11}) + \lambda_1(B_{22})$.

Proof Since B is positive semidefinite, there exists an $n \times n$ matrix C such that $B = CC'$. Partition $C = \begin{bmatrix} C_1 \\ C_2 \end{bmatrix}$ so that

$$
B = \begin{bmatrix} B_{11} & B_{12} \\ B_{21} & B_{22} \end{bmatrix} = \begin{bmatrix} C_1C_1' & C_1C_2' \\ C_2C_1' & C_2C_2' \end{bmatrix}.
$$

Now

$$
\begin{aligned}
\lambda_1(B) &= \lambda_1(CC') \\
&= \lambda_1(C'C) \\
&= \lambda_1(C_1'C_1 + C_2'C_2) \\
&\leq \lambda_1(C_1'C_1) + \lambda_1(C_2'C_2) \qquad \text{by Lemma 3.19} \\
&= \lambda_1(C_1C_1') + \lambda_1(C_2C_2') \\
&= \lambda_1(B_{11}) + \lambda_1(B_{22}),
\end{aligned}
$$

and the proof is complete.   □

Lemma 3.21 *Let B be an $n \times n$ symmetric matrix and suppose B is partitioned as*

$$
B = \begin{bmatrix} B_{11} & B_{12} \\ B_{21} & B_{22} \end{bmatrix},
$$

where B_{11} is $p \times p$. Then

$$
\lambda_1(B) + \lambda_n(B) \leq \lambda_1(B_{11}) + \lambda_1(B_{22}).
$$

Proof We have

$$B - \lambda_n(B)I_n = \begin{bmatrix} B_{11} - \lambda_n(B)I_p & B_{12} \\ B_{21} & B_{22} - \lambda_n(B)I_{n-p} \end{bmatrix}.$$

Since $B - \lambda_n(B)I_n$ is positive semidefinite, by Lemma 3.20 we get

$$\lambda_1(B - \lambda_n(B)I_n) \leq \lambda_1(B_{11} - \lambda_n(B)I_p) + \lambda_1(B_{22} - \lambda_n(B)I_{n-p}).$$

Therefore,

$$\lambda_1(B) - \lambda_n(B) \leq \lambda_1(B_{11}) - \lambda_n(B) + \lambda_1(B_{22}) - \lambda_n(B),$$

and hence

$$\lambda_1(B) + \lambda_n(B) \leq \lambda_1(B_{11}) + \lambda_1(B_{22}).$$

This completes the proof. □

Lemma 3.22 *Let B be a symmetric matrix partitioned as*

$$B = \begin{bmatrix} 0 & B_{12} & \cdots & B_{1k} \\ B_{21} & 0 & \cdots & B_{2k} \\ \vdots & \vdots & \ddots & \vdots \\ B_{k1} & B_{k2} & \cdots & 0 \end{bmatrix}.$$

Then $\lambda_1(B) + (k-1)\lambda_n(B) \leq 0$.

Proof We prove the result by induction on k. When $k = 2$ the result follows by Lemma 3.21. So assume the result to be true for $k - 1$. Let C be the principal submatrix of B obtained by deleting the last row and column of blocks. If $\lambda_{\min}(C)$ denotes the minimum eigenvalue of C, then by the induction assumption,

$$\lambda_1(C) + (k-2)\lambda_{\min}(C) \leq 0. \tag{3.4}$$

By Lemma 3.21,

$$\lambda_1(B) + \lambda_n(B) \leq \lambda_1(C). \tag{3.5}$$

Since the minimum eigenvalue of a symmetric matrix does not exceed that of a principal submatrix,

$$\lambda_n(B) \leq \lambda_{\min}(C). \tag{3.6}$$

From (3.4) and (3.5) we get

$$\lambda_1(B) + \lambda_n(B) + (k-2)\lambda_{\min}(C) \leq 0. \tag{3.7}$$

Using (3.6) and (3.7) we have

$$\lambda_1(B) + (k-1)\lambda_n(B) \le 0$$

and the proof is complete.   □

We are now ready to prove the following bound due to Hoffman.

Theorem 3.23 *Let G be a graph with n vertices and with at least one edge. Then*

$$\chi(G) \ge 1 - \frac{\lambda_1(G)}{\lambda_n(G)}.$$

Proof Let A be the adjacency matrix of G. If $\chi(G) = k$, then after a relabeling of the vertices of G we may write

$$A = \begin{bmatrix} 0 & A_{12} & \cdots & A_{1k} \\ A_{21} & 0 & \cdots & A_{2k} \\ \vdots & \vdots & \ddots & \vdots \\ A_{k1} & A_{k2} & \cdots & 0 \end{bmatrix}.$$

By Lemma 3.22,

$$\lambda_1(A) + (k-1)\lambda_n(A) \le 0. \tag{3.8}$$

If G has at least one edge then the eigenvalues of G are not all equal to zero, and $\lambda_n(A) < 0$. Thus, from (3.8),

$$\chi(G) = k \ge 1 - \frac{\lambda_1(A)}{\lambda_n(A)} = 1 - \frac{\lambda_1(G)}{\lambda_n(G)}.$$

This completes the proof.   □

3.4 Energy of a Graph

An interesting quantity in Hückel theory is the sum of the energies of all the electrons in a molecule, the so-called total π-electron energy E_π. For a molecule with $n = 2k$ atoms, the total π-electron energy can be shown to be $E_\pi = 2\sum_{i=1}^{k} \lambda_i$, where λ_i, $i = 1, 2, \ldots, k$, are the k largest eigenvalues of the adjacency matrix of the graph of the molecule. For a bipartite graph, because of the symmetry of the spectrum, we can write $E_\pi = \sum_{i=1}^{n} |\lambda_i|$, and this has motivated the following definition.

For any (not necessarily bipartite) graph G, the *energy* of the graph is defined as $\varepsilon(G) = \sum_{i=1}^{n} |\lambda_i|$, where $\lambda_1, \ldots, \lambda_n$ are the eigenvalues of the adjacency matrix of G.

Characterizing the set of positive numbers that can occur as energy of a graph has been a problem of interest. We now prove that the energy can never be an odd integer. In fact, we show that if the energy is rational then it must be an even integer. Some inequalities for energy and a characterization of graphs with maximum energy will be treated in a later section.

We need some preliminaries. Let A and B be matrices of order $m \times n$ and $p \times q$, respectively. The Krönecker product of A and B, denoted $A \otimes B$, is the $mp \times nq$ block matrix $[a_{ij} B]$. It can be verified from the definition that

$$(A \otimes B)(C \otimes D) = AC \otimes BD. \tag{3.9}$$

Several important properties of the Kronecker product are consequences of (3.9). The next result, although proved for symmetric matrices, is also true in general.

Lemma 3.24 *Let A and B be symmetric matrices of order $m \times m$ and $n \times n$, respectively. If $\lambda_1, \ldots, \lambda_m$ and $\mu_1, \ldots, \mu_n$ are the eigenvalues of A and B, respectively, then the eigenvalues of $A \otimes I_n + I_m \otimes B$ are given by $\lambda_i + \mu_j$; $i = 1, \ldots, m$; $j = 1, \ldots, n$.*

Proof Let P and Q be orthogonal matrices such that

$$P'AP = \text{diag}(\lambda_1, \ldots, \lambda_m), \quad Q'BQ = \text{diag}(\mu_1, \ldots, \mu_n).$$

Then by (3.9),

$$(P \otimes Q)(A \otimes I_n + I_m \otimes B)(P' \otimes Q') = PAP' \otimes QQ' + PP' \otimes QBQ'$$
$$= \text{diag}(\lambda_1, \ldots, \lambda_m) \otimes I_n + I_m \otimes \text{diag}(\mu_1, \ldots, \mu_n).$$

The proof is complete in view of the fact that $\text{diag}(\lambda_1, \ldots, \lambda_m) \otimes I_n + I_m \otimes \text{diag}(\mu_1, \ldots, \mu_n)$ is a diagonal matrix with $\lambda_i + \mu_j$; $i = 1, \ldots, m$; $j = 1, \ldots, n$, on the diagonal. $\qquad\square$

The following result is similarly proved.

Lemma 3.25 *Let A and B be symmetric matrices of order $m \times m$ and $n \times n$, respectively. If $\lambda_1, \ldots, \lambda_m$ and $\mu_1, \ldots, \mu_n$ are the eigenvalues of A and B, respectively, then the eigenvalues of $A \otimes B$ are given by $\lambda_i \mu_j$; $i = 1, \ldots, m$; $j = 1, \ldots, n$.*

Let G and H be graphs with vertex sets $V(G)$ and $V(H)$, respectively. The Cartesian product of G and H, denoted by $G \times H$, is the graph defined as follows. The vertex set of $G \times H$ is $V(G) \times V(H)$. The vertices (u, v) and (u', v') are adjacent if either $u = u'$ and v is adjacent to v' in H, or $v = v'$ and u is adjacent to u' in G.

Let $|V(G)| = m$, $|V(H)| = n$, and suppose A and B are the adjacency matrices of G and H, respectively. It can be verified that the adjacency matrix of $G \times H$ is $A \otimes I_n + I_m \otimes B$. The following result follows from this observation and by Lemma 3.24.

Lemma 3.26 *Let G and H be graphs with m and n vertices, respectively. If $\lambda_1, \ldots, \lambda_m$ and $\mu_1, \ldots, \mu_n$ are the eigenvalues of G and H, respectively, then the eigenvalues of $G \times H$ are given by $\lambda_i + \mu_j$, $i = 1, \ldots, m$; $j = 1, \ldots, n$.*

We are now in a position to prove the next result, which identifies the possible values that the energy of a graph can attain, among the set of rational numbers.

Theorem 3.27 *Let G be a graph with n vertices. If the energy $\varepsilon(G)$ of G is a rational number then it must be an even integer.*

Proof Let $\lambda_1, \ldots, \lambda_k$ be the positive eigenvalues of G. The trace of the adjacency matrix is zero, and hence the sum of the positive eigenvalues of G equals the sum of the absolute values of the negative eigenvalues of G. It follows from the definition of energy that $\varepsilon(G) = 2(\lambda_1 + \cdots + \lambda_k)$. Note that by Lemma 3.26, $\lambda_1 + \cdots + \lambda_k$ is an eigenvalue of $G \times \cdots \times G$, taken k times. The characteristic polynomial of the adjacency matrix is a monic polynomial with integer coefficients, and a rational root of such a polynomial must be an integer. Thus, if $\lambda_1 + \cdots + \lambda_k$ is rational, then it must be an integer. It follows that if $\varepsilon(G)$ is rational, then it must be an even integer. □

3.5 Antiadjacency Matrix of a Directed Graph

We consider directed graphs in this section. Let G be a directed graph with $V(G) = \{1, \ldots, n\}$. The adjacency matrix A of G is defined in a natural way. Thus, the rows and the columns of A are indexed by $V(G)$. For $i \neq j$, if there is an edge from i to j, then $a_{ij} = 1$, otherwise $a_{ij} = 0$. We set $a_{ii} = 0$, $i = 1, \ldots, n$. The matrix $B = J - A$ will be called the *antiadjacency matrix* of G. Recall that a Hamiltonian path is a path meeting all the vertices in the graph. It turns out that if G is acyclic, i.e., has no directed cycles, then det $B = 1$ if G has a directed Hamiltonian path, otherwise det $B = 0$. We will prove a result that is more general. First we prove a preliminary result.

Lemma 3.28 *Let B be a $0 - 1$ $n \times n$ matrix such that $b_{ij} = 1$ if $i \geq j$. Then det B equals 1 if $b_{12} = b_{23} = \cdots = b_{n-1n} = 0$; otherwise det $B = 0$.*

Proof If $b_{12} = 1$ then the first two columns of B have all entries equal to 1, and hence det $B = 0$. So let $b_{12} = 0$. In B subtract the second column from the first column. Then the first column has all entries equal to 0, except the $(1, 1)$-entry, which equals 1. Expand the determinant along the first column and use induction on n to complete the proof. □

Corollary 3.29 *Let G be a directed, acyclic graph with $V(G) = \{1, \ldots, n\}$. Let B be the antiadjacency matrix of G. Then det $B = 1$ if G has a Hamiltonian path, and det $B = 0$, otherwise.*

Proof Suppose G has a Hamiltonian path, and without loss of generality, let it be $1 \rightarrow 2 \rightarrow \cdots \rightarrow n$. Since G is acyclic, there cannot be an edge from i to j for $i > j$ and hence $b_{ij} = 1$, $i \geq j$. Also, $b_{12} = \cdots = b_{n-1,n} = 0$, and by Lemma 3.28 $\det B = 1$.

Conversely, suppose G has no Hamiltonian path. Since G is acyclic, there must be a vertex of G which is a source, i.e., a vertex of in-degree 0. Without loss of generality, let it be 1. In $G \setminus \{1\}$ there is a source, which we assume to be 2. Continuing this way, let i be the source in $G \setminus \{1, \ldots, i-1\}$, $i = 2, \ldots, n$. Then there is no edge from j to i, $j > i$, and hence B has ones on and below the main diagonal. Since G has no Hamiltonian path, there must exist i in $\{1, \ldots, n-1\}$ such that $b_{i,i+1} = 1$. Then by Lemma 3.28 $\det B = 0$. $\qquad\square$

Theorem 3.30 *Let G be a directed, acyclic graph with $V(G) = \{1, \ldots, n\}$. Let B be the antiadjacency matrix of G, and let*

$$\det(xB + I) = \sum_{i=0}^{n} c_i x^i.$$

Then $c_0 = 1$ and c_i equals the number of directed paths of i vertices in G, $i = 1, \ldots, n$.

Proof By expanding the determinant it can be seen that the coefficient of x^i in $\det(xB + I)$ is the sum of the principal minors of B of order i, $i = 1, \ldots, n$. By Corollary 3.29, a principal minor of B is 1 if and only if the subgraph induced by the corresponding vertices has a Hamiltonian path. Note that this induced subgraph cannot have another Hamiltonian path, otherwise it will contain a cycle. Thus, the sum of the nonsingular $i \times i$ principal minors of B equals the number of paths in G of i vertices. This completes the proof. $\qquad\square$

Example 3.31 Consider the acyclic directed graph G:

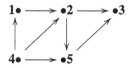

The antiadjacency matrix of G is given by

$$B = \begin{bmatrix} 1 & 0 & 1 & 1 & 1 \\ 1 & 1 & 0 & 1 & 0 \\ 1 & 1 & 1 & 1 & 1 \\ 0 & 0 & 1 & 1 & 0 \\ 1 & 1 & 0 & 1 & 1 \end{bmatrix}.$$

It can be checked that $\det(xB + I) = x^5 + 4x^4 + 7x^3 + 7x^2 + 5x + 1$. The directed paths of G are listed Table 3.1, according to the number of vertices in the path:

Thus, the coefficient of x^i in $\det(xB + I)$ equals the number of directed paths of i vertices, $i = 1, \ldots, 5$.

3.6 Nonsingular Trees

The adjacency matrix of a tree may or may not be nonsingular. For example, the adjacency matrix of a path is nonsingular if and only if the path has an even number of vertices. We say that a tree is nonsingular if its adjacency matrix is nonsingular. A matching in a graph is a set of edges, no two of which have a common vertex. A matching is perfect if every vertex in the graph is incident to an edge in the matching. A simple criterion for a tree to be nonsingular is given in the next result.

Lemma 3.32 *Let T be a tree with $V(T) = \{1, \ldots, n\}$, and let A be the adjacency matrix of T. Then A is nonsingular if and only if T has a perfect matching.*

Proof Using the notation of Theorem 3.8,

$$\det A = \sum (-1)^{n - c_1(H) - c(H)} 2^{c(H)},$$

where the summation is over all spanning elementary subgraphs H of T. If $\det A$ is nonzero then T has an elementary spanning subgraph. In the case of a tree, $c(H) = 0$ for any H. Thus, an elementary spanning subgraph consists exclusively of edges, which clearly must form a perfect matching.

To prove the converse, first observe that if T has a perfect matching then it must be unique. This statement is easily proved by induction on the number of vertices. Thus, if T has perfect matching then only one nonzero term is obtained in the above summation and hence $\det A$ must be nonzero. This completes the proof.   □

Our next objective is to provide a formula for the inverse of the adjacency matrix of a nonsingular tree. If i and j are vertices of the tree then we denote by $P(i, j)$ the unique ij-path in the tree. The length of $P(i, j)$ is $d(i, j)$, the distance between i and j.

Table 3.1 Directed paths in the graph G

Number of vertices	Directed path(s)
5	41253
4	1253, 4125, 4123, 4253
3	123, 125, 412, 423, 425, 453, 253
2	12, 23, 25, 41, 42, 45, 53
1	1, 2, 3, 4, 5

If T has a perfect matching $\mathcal{M}$, then $P(i, j)$ will be called an *alternating path* if its edges are alternately in $\mathcal{M}$ and $\mathcal{M}^c$, the first edge and the last edge being in $\mathcal{M}$. If $P(i, j)$ has only one edge and that edge is in $\mathcal{M}$, then $P(i, j)$ is also considered to be alternating. We note that if $P(i, j)$ is alternating then $d(i, j)$ must be odd.

Theorem 3.33 *Let T be a nonsingular tree with $V(T) = \{1, \ldots, n\}$ and let A be the adjacency matrix of T. Let $\mathcal{M}$ be the perfect matching in T. Let $B = [b_{ij}]$ be the $n \times n$ matrix defined as follows: $b_{ij} = 0$ if $i = j$ or if $P(i, j)$ is not alternating. If $P(i, j)$ is alternating, then set*

$$b_{ij} = (-1)^{\frac{d(i.j)-1}{2}}.$$

Then $B = A^{-1}$.

Proof We assume, without loss of generality, that 1 is adjacent to $2, \ldots, k$, and that the edge $\{1, 2\} \in \mathcal{M}$. Since $a_{1j} = 0, j > k$, then $a_{1j}b_{j1} = 0$, if $j > k$. For $j = 3, \ldots, k$, $P(j, 1)$ is not alternating and hence $a_{1j}b_{j1} = 0$ for these values of j. Finally, $a_{12}b_{21} = 1$, since $a_{12} = 1$ and $P(1, 2)$ is alternating. Combining these observations we see that

$$\sum_{j=1}^{n} a_{1j}b_{j1} = 1. \tag{3.10}$$

Let T_i be the component of $T \setminus \{1\}$, containing i, $i = 2, \ldots, k$. If $\ell \in V(T_2)$ then there is no alternating path from ℓ to j, $j = 2, \ldots, k$, and hence

$$\sum_{j=1}^{n} a_{1j}b_{j\ell} = 0. \tag{3.11}$$

Now suppose $\ell \in V(T_i)$ for some $i \in \{3, \ldots, k\}$. Note that $P(\ell, j)$ is not alternating if $j \in \{3, \ldots, k\} \setminus \{i\}$. Also, if $P(\ell, i)$ is alternating then $P(\ell, 2)$ is alternating as well, and furthermore, $d(\ell, i) = d(\ell, 2) - 2$. Thus, $b_{\ell j} = 0$, $j \in \{3, \ldots, k\} \setminus \{i\}$, and $b_{2\ell} + b_{i\ell} = 0$. It follows that

$$\sum_{j=1}^{n} a_{1j}b_{j\ell} = 0. \tag{3.12}$$

From (3.10), (3.11) and (3.12) it follows that the first row of AB is the same as the first row of I. We can similarly show that any row of AB is the same as the corresponding row of I and hence $B = A^{-1}$. $\square$

A signature matrix is a diagonal matrix with ± 1 on the diagonal. A pendant vertex is a vertex of degree 1.

Theorem 3.34 *Let T be a nonsingular tree with $V(T) = \{1, \ldots, n\}$ and let A be the adjacency matrix of T. Then there exists a signature matrix F such that $FA^{-1}F$ is the adjacency matrix of a graph.*

Proof We assume, without loss of generality, that 1 is a pendant vertex of T. By Lemma 3.32, T has a perfect matching, which we denote by $\mathcal{M}$. For $i = 1, \ldots, n$, let n_i be the number of edges in $P(1, i)$ that are not in $\mathcal{M}$. (We set $n_1 = 0$.) Let $f_i = (-1)^{n_i}$, $i = 1, \ldots, n$, and let $F = \text{diag}(f_1, \ldots, f_n)$. Let $B = A^{-1}$ and note that a formula for B is given in Theorem 3.33. The (i, j)-element of FBF is $f_i f_j b_{ij}$, which equals 0 if and only if $b_{ij} = 0$.

Let $i, j \in V(T)$ and suppose $b_{ij} \neq 0$. By Theorem 3.33, $P(i, j)$ is an alternating path. Let k be the vertex in $P(i, j)$ that is nearest to 1. Let $r = \frac{d(i,j)-1}{2}$, which is the number of edges in $P(i, j)$ that are not in $\mathcal{M}$. It can be verified that

$$n_i + n_j - 2n_k = r. \tag{3.13}$$

It follows by (3.13) and Theorem 3.33 that

$$f_i f_j b_{ij} = (-1)^{n_i}(-1)^{n_j}(-1)^r = (-1)^{2n_k} = 1.$$

Thus, each entry of FBF is either 0 or 1, and clearly, FBF is symmetric. Hence, FBF is the adjacency matrix of a graph. $\square$

The inverse of the nonsingular tree T will be defined as the graph with adjacency matrix $FA^{-1}F$ as given in Theorem 3.33. We denote the inverse of T by T^{-1}.

Example 3.35 Consider the tree T as shown. Edges in the perfect matching are shown as dashed lines.

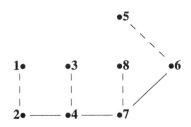

The graph T^{-1} is as follows:

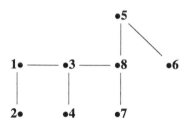

In Example 3.35 it turns out that T^{-1} is a tree as well, though this is not always the case. If T has an alternating path of length at least 5, then it will result in a cycle in T^{-1}, as can be seen from Theorem 3.33. We now proceed to identify conditions under which T^{-1} is a tree.

Let T be a nonsingular tree with adjacency matrix A. The adjacency matrix of T^{-1} is obtained by taking A^{-1} and replacing each entry by its absolute value.

Lemma 3.36 *Let T be a nonsingular tree with $V(T) = \{1, \ldots, n\}$. Then T^{-1} is a connected graph.*

Proof Let A and B be the adjacency matrices of T and T^{-1}, respectively. If T^{-1} is disconnected then, after a relabeling of vertices,

$$B = \begin{bmatrix} B_{11} & 0 \\ 0 & B_{22} \end{bmatrix}.$$

Since B and A^{-1} have the same pattern of zero-nonzero entries, A must also be a direct sum of two matrices. This is a contradiction, as T is connected, and the proof is complete.   □

Corollary 3.37 *Let T be a nonsingular tree with $V(T) = \{1, \ldots, n\}$. Then the number of alternating paths in T, which equals the number of edges in T^{-1}, is at least $n - 1$.*

Proof Let A be the adjacency matrix of T. As seen in the proof of Theorem 3.33, each alternating path in T corresponds to a nonzero entry in A^{-1}, which in turn, corresponds to an edge in T^{-1}. By Lemma 3.36, T^{-1} is connected and hence it has at least $n - 1$ edges.   □

A *corona tree* is a tree obtained by attaching a new pendant vertex to each vertex of a given tree.

Theorem 3.38 *Let T be a nonsingular tree with $V(T) = \{1, \ldots, 2n\}$. Then the following conditions are equivalent:*

(i) the number of alternating paths in T has the minimum possible value $2n - 1$;
(ii) T^{-1} is a tree;
(iii) T is a corona tree;
(iv) T^{-1} is isomorphic to T.

Proof $(i) \Rightarrow (ii)$: As remarked earlier, the number of alternating paths in T equals the number of edges in T^{-1}. If there are $2n - 1$ alternating paths in T, then T^{-1} has $2n - 1$ edges, and since by Lemma 3.36, T^{-1} is a connected graph, it follows that T^{-1} is a tree.

$(ii) \Rightarrow (iii)$: Suppose T^{-1} is a tree. If T has 4 vertices, then T must be the path on 4 vertices, and it can be verified that T^{-1} is also the path on 4 vertices. Therefore, we may assume that T has at least 6 vertices. Let $\mathcal{M}$ be the perfect matching in T, and we assume that the edges in $\mathcal{M}$ are $\{u_i, v_i\}$, $i = 1, \ldots, n$. We claim that for any edge $\{u_i, v_i\}$ in $\mathcal{M}$, either u_i or v_i is a pendant vertex. Otherwise, u_i must be adjacent to a vertex other than v_i, say u_j, while v_i must be adjacent to a vertex other than u_i, say v_k. Then $v_j - u_j - u_i - v_i - v_k - u_k$ is an alternating path of length 5 in T, in which case T^{-1} cannot be a tree. Thus, one of the vertices of $\{u_i, v_i\}$ is a pendant vertex for each $i = 1, \ldots, n$. It follows that T is a corona tree.

$(iii) \Rightarrow (iv)$: Let T be a corona tree and assume, without loss of generality, that vertices $n + 1, \ldots, 2n$ are pendant. Let B be the adjacency matrix of the subtree induced by $\{1, \ldots, n\}$. Then the adjacency matrix A of T has the form

$$A = \begin{bmatrix} B & I \\ I & 0 \end{bmatrix}.$$

Then

$$A^{-1} = \begin{bmatrix} 0 & I \\ I & -B \end{bmatrix}.$$

Therefore, the adjacency matrix of T^{-1} is

$$\begin{bmatrix} 0 & I \\ I & B \end{bmatrix}.$$

It follows that T and T^{-1} are isomorphic and the proof is complete.

$(iv) \Rightarrow (i)$: If T^{-1} is isomorphic to T then it must have $2n - 1$ edges. Since T^{-1} is connected by Lemma 3.36, T^{-1} must be a tree. $\qquad\square$

Exercises

1. Verify that the following two graphs are nonisomorphic, yet they have the same eigenvalues.

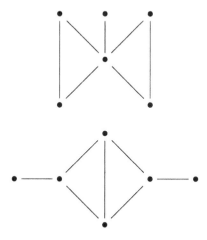

2. List the spanning elementary subgraphs of K_4. Hence, using Theorem 3.8, show that $\det A(K_4) = -3$.

3. Using the notation of Theorem 3.10, show that c_4 is equal to the number of pairs of disjoint edges minus twice the number of 4-cycles in G.

4. Let G be a planar graph with n vertices. Show that $\lambda_1(G) \leq -3\lambda_n(G)$.

5. Determine the energies of K_n and $K_{m,n}$. Conclude that any even positive integer is the energy of a graph.

6. Let G and H be graphs with vertex sets $V(G)$ and $V(H)$, respectively. The tensor product of G and H, denoted $G \otimes H$, is the graph with vertex set $V(G) \times V(H)$, and two vertices (u, v) and (u', v') are adjacent if and only if u, u' are adjacent in G, and v, v' are adjacent in H. Show that if A and B are the adjacency matrices of G and H, respectively, then $A \otimes B$ is the adjacency matrix of $G \otimes H$. Hence, show that $\varepsilon(G \otimes H) = \varepsilon(G)\varepsilon(H)$.

7. Let G be a graph with at least one edge. Show that the graphs $G \otimes K_2 \otimes K_2$ and $G \otimes C_4$ have the same energy, though they are not isomorphic.

8. Let G be a graph with n vertices and let A be the adjacency matrix of G. Let G_1 and G_2 be the graphs with $2n$ vertices with adjacency matrices $\begin{bmatrix} 0 & A \\ A & 0 \end{bmatrix}$ and $\begin{bmatrix} A & A \\ A & A \end{bmatrix}$, respectively. Show that G_1 and G_2 have the same energy.

9. Let T be a tree with $V(T) = \{1, \ldots, n\}$, and let A be the adjacency matrix of T. Show that A is totally unimodular.

Much of the basic material covered in this chapter can be found in [SW78]. Other relevant references are: Sect. 3.4: [BP04], [IV06], Sect. 3.5: [ST96], Sect. 3.6: [BNP06]. Theorem 3.33 can be found in [BDH88]. Exercises 6 and 7 are from [BA04]. For a wealth of information on the spectrum of the adjacency matrix, see [CDS95].

References and Further Readings

[BA04]   Balakrishnan, R.: The energy of a graph. Linear Algebra Appl. **387**, 287–295 (2004)
[BP04]   Bapat, R.B., Pati, S.: Energy of a graph is never an odd integer. Bull. Kerala Math. Assoc. **1**(2), 129–132 (2004)
[BNP06] Barik, S., Neumann, M., Pati, S.: On nonsingular trees and a reciprocal eigenvalue property. Linear Multilinear Algebra **54**(6), 453–465 (2006)
[BDH88] Buckley, F., Doty, L.L., Harary, F.: On graphs with signed inverses. Networks **18**(3), 151–157 (1988)
[CDS95] Cvetković, D.M., Doob, M., Sachs, H.: Spectra of Graphs, Theory and Applications, 3rd edn. Johann Ambrosius Barth, Heidelberg (1995)
[IV06]   Indulal, G., Vijayakumar, A.: On a pair of equienergetic graphs. MATCH Commun. Math. Comput. Chem. **55**(1), 83–90 (2006)
[SW78]   Schwenk, A.J., Wilson, R.J.: On the eigenvalues of a graph. In: Beineke, L.W., Wilson, R.J. (eds.) Selected Topics in Graph Theory, pp. 307–336. Academic Press, New York (1978)
[ST96]   Stanley, R.: A matrix for counting paths in acyclic digraphs. J. Comb. Theor. Ser. A **74**, 169–172 (1996)

Chapter 4
Laplacian Matrix

Let G be a graph with $V(G) = \{1, \ldots, n\}$ and $E(G) = \{e_1, \ldots, e_m\}$. The Laplacian matrix of G, denoted by $L(G)$, is the $n \times n$ matrix defined as follows. The rows and columns of $L(G)$ are indexed by $V(G)$. If $i \neq j$ then the (i, j)-entry of $L(G)$ is 0 if vertex i and j are not adjacent, and it is -1 if i and j are adjacent. The (i, i)-entry of $L(G)$ is d_i, the degree of the vertex i, $i = 1, 2, \ldots, n$.

Let $D(G)$ be the diagonal matrix of vertex degrees. If $A(G)$ is the adjacency matrix of G, then note that $L(G) = D(G) - A(G)$.

Suppose each edge of G is assigned an orientation, which is arbitrary but fixed. Let $Q(G)$ be the incidence matrix of G. Then observe that $L(G) = Q(G)Q(G)'$. This can be seen as follows. The rows of $Q(G)$ are indexed by $V(G)$. The (i, j)-entry of $Q(G)Q(G)'$ is the inner product of the rows i and j of $Q(G)$. If $i = j$ then the inner product is clearly d_i, the degree of the vertex i. If $i \neq j$, then the inner product is -1 if the vertices i and j are adjacent, and zero otherwise.

The identity $L(G) = Q(G)Q(G)'$ suggests that the Laplacian might depend on the orientation, although it is evident from the definition that the Laplacian is independent of the orientation.

Example 4.1 Consider the graph

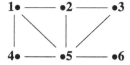

Its Laplacian matrix is given by

$$
L(G) = \begin{bmatrix}
3 & -1 & 0 & -1 & -1 & 0 \\
-1 & 3 & -1 & 0 & -1 & 0 \\
0 & -1 & 2 & 0 & -1 & 0 \\
-1 & 0 & 0 & 2 & -1 & 0 \\
-1 & -1 & -1 & -1 & 5 & -1 \\
0 & 0 & 0 & 0 & -1 & 1
\end{bmatrix}.
$$

© Springer-Verlag London 2014
R.B. Bapat, *Graphs and Matrices*, Universitext,
DOI 10.1007/978-1-4471-6569-9_4

4.1 Basic Properties

We begin with a preparatory result.

Lemma 4.2 *Let X be an $n \times n$ matrix with zero row and column sums. Then the cofactors of any two elements of X are equal.*

Proof As usual, let $X(i|j)$ denote the matrix obtained by deleting row i and column j of X. In $X(1|1)$ add all the columns to the first column. Then the first column of $X(1|1)$ becomes the negative of $[x_{21}, \ldots, x_{n1}]'$, in view of the fact that the row sums of X are zero. Thus, we conclude that $\det X(1|1) = -\det X(1|2)$. In other words, the cofactors of x_{11} and x_{12} are equal. A similar argument shows that the cofactor of x_{ij} equals that of x_{ik}, for any i, j, k.

Now using the fact that the column sums of X are zero, we conclude that the cofactor of x_{ij} equals that of x_{kj}, for any i, j, k. It follows that the cofactors of any two elements of X are equal.   □

Some elementary properties of the Laplacian are summarized in the next result.

Lemma 4.3 *Let G be a graph with $V(G) = \{1, \ldots, n\}$ and $E(G) = \{e_1, \ldots, e_m\}$. Then the following assertions hold.*

(i) *$L(G)$ is a symmetric, positive semidefinite matrix.*
(ii) *The rank of $L(G)$ equals $n - k$, where k is the number of connected components of G.*
(iii) *For any vector x,*
$$x'L(G)x = \sum_{i \sim j} (x_i - x_j)^2.$$

(iv) *The row and the column sums of $L(G)$ are zero.*
(v) *The cofactors of any two elements of $L(G)$ are equal.*

Proof (i) It is obvious from $L(G) = Q(G)Q(G)'$ that $L(G)$ is symmetric and positive semidefinite.

(ii) This follows from the fact that

$$\operatorname{rank} L(G) = \operatorname{rank} Q(G)Q(G)' = \operatorname{rank} Q(G)$$

and by using Theorem 2.3.

(iii) Note that $x'L(G)x = x'Q(G)Q(G)'x$. The vector $x'Q(G)$ is indexed by $E(G)$. In fact, the entry of $x'Q(G)$, indexed by the edge $e = \{i, j\}$, is $x_i - x_j$. Hence the result follows.

(iv) This follows from the definition $L(G) = D(G) - A(G)$.

(v) This is evident from Lemma 4.2 and (iv).   □

4.2 Computing Laplacian Eigenvalues

Recall that J denotes the square matrix with all entries equal to 1, and the order of the matrix will be clear from the context.

Lemma 4.4 *The eigenvalues of the $n \times n$ matrix $aI + bJ$ are a with multiplicity $n - 1$, and $a + nb$ with multiplicity 1.*

Proof As observed in the proof of Theorem 3.4, the eigenvalues of J are 0 with multiplicity $n - 1$, and n with multiplicity 1. It follows that the eigenvalues of bJ are 0 with multiplicity $n - 1$, and nb with multiplicity 1. Then the eigenvalues of $aI + bJ$ must be a with multiplicity $n - 1$, and $a + nb$ with multiplicity 1. $\square$

It follows from Lemma 4.4 that $L(K_n) = nI - J$ has eigenvalues n with multiplicity $n - 1$, and 0 with multiplicity 1. The following result is often useful in calculating the eigenvalues of Laplacians.

Lemma 4.5 *Let G be a graph with $V(G) = \{1, 2, \ldots, n\}$. Let the eigenvalues of $L(G)$ be $\lambda_1 \geq \cdots \geq \lambda_{n-1} \geq \lambda_n = 0$. Then the eigenvalues of $L + aJ$ are $\lambda_1 \geq \cdots \geq \lambda_{n-1}$ and na.*

Proof There exists an orthogonal matrix P whose columns form eigenvectors of $L(G)$. We assume that the last column of P is the vector with each component $\frac{1}{\sqrt{n}}$; this being an eigenvector for the eigenvalue 0. Then $P'L(G)P = \text{diag}(\lambda_1, \ldots, \lambda_n)$. Note that any column of P other than the last column is orthogonal to the last column, and hence

$$
JP = \begin{bmatrix} 0 \cdots 0 & \sqrt{n} \\ 0 \cdots 0 & \sqrt{n} \\ & \vdots \\ 0 \cdots 0 & \sqrt{n} \end{bmatrix}.
$$

It follows that $P'JP = \text{diag}(0, \ldots, 0, n)$. Therefore,

$$
P'(L(G) + aJ)P = \text{diag}(\lambda_1, \ldots, \lambda_{n-1}, na)
$$

and hence the eigenvalues of $L(G) + aJ$ must be as asserted. $\square$

An application of Lemma 4.5 is illustrated in the next result.

Lemma 4.6 *Let G be the graph obtained by removing p disjoint edges from K_n, $n \geq 2p$. Then the eigenvalues of $L(G)$ are $n - 2$ with multiplicity p, n with multiplicity $n - p - 1$, and 0 with multiplicity 1.*

Proof Assume, without loss of generality, that the edges

$$
\{1, 2\}, \{3, 4\}, \ldots, \{2p - 1, 2p\}
$$

are removed from K_n to obtain G. Then $L(G) + J$ is a block diagonal matrix, in which the block $\begin{bmatrix} n-1 & 1 \\ 1 & n-1 \end{bmatrix}$ appears p times, and nI_{n-2p} appears once. Therefore, the eigenvalues of $L(G) + J$ are $n-2$ with multiplicity p, and n with multiplicity $n-p$. It follows by Lemma 4.5 that the eigenvalues of $L(G)$ are $n-2$ with multiplicity p, n with multiplicity $n-p-1$, and 0 with multiplicity 1. $\square$

4.3 Matrix-Tree Theorem

Theorem 4.7 *Let G be a graph with $V(G) = \{1, 2, \ldots, n\}$. Let W be a nonempty proper subset of $V(G)$. Then the determinant of $L(W|W)$ equals the number of spanning forests of G with $|W|$ components in which each component contains one vertex of W.*

Proof Assign an orientation to each edge of G and let Q be the incidence matrix. We assume, without loss of generality, that $W = \{1, 2, \ldots, k\}$.

By the Cauchy–Binet formula,

$$\det L(W|W) = \sum (\det Q[W^c|Z])^2,$$

where the summation is over all $Z \subset E(G)$ with $|Z| = n - k$.

Since by Lemma 2.6 Q is totally unimodular, then $(\det Q[W^c|Z])^2$ equals 0 or 1. Thus, $\det L(W|W)$ equals the number of nonsingular submatrices of Q with row set W^c.

In view of the discussion in Sect. 2.2, $Q[W^c|Z]$ is nonsingular if and only if each component of the corresponding substructure is a rootless tree. Hence, there is a one-to-one correspondence between nonsingular submatrices of Q with row set W^c and spanning forests of G with $|W|$ components in which each component contains one vertex of W. $\square$

The following result, which is an immediate consequence of Lemma 4.3 and Theorem 4.7, is the well-known matrix-tree theorem.

Theorem 4.8 *Let G be a graph with $V(G) = \{1, 2, \ldots, n\}$. Then the cofactor of any element of $L(G)$ equals the number of spanning trees of G.*

Proof Setting $W = \{1\}$ in Theorem 4.7, it follows that $\det L(1|1)$ equals the number of spanning forests of G with one component, which is the same as the number of spanning trees of G. By Lemma 4.3 all the cofactors of $L(G)$ are equal and the result is proved. $\square$

We remark that in Theorem 4.8 it is not necessary to assume that G is connected. For, if G is disconnected then it has no spanning trees. At the same time, the rank of $L(G)$ is at most $n-2$ and hence all its cofactors are zero.

The wheel W_n is a graph consisting of a cycle on n vertices, $1, 2, \ldots, n$, and the vertex $n + 1$, which is adjacent to each of $1, 2, \ldots, n$. The wheel W_6 is shown in the figure.

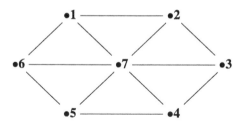

Let C_n denote the cycle on n vertices.

Lemma 4.9 *The eigenvalues of $L(C_n)$ are $2 - 2\cos\frac{2\pi j}{n}$, $j = 1, \ldots, n$.*

Proof By Theorem 3.6, the eigenvalues of $A(C_n)$ are $2\cos\frac{2\pi j}{n}$, $j = 1, \ldots, n$. Since $L(C_n) = 2I - A(C_n)$, the result follows. $\qquad\square$

If we delete row $n + 1$ and column $n + 1$ from $L(W_n)$, we obtain the matrix $L(C_n) + I_n$. By Lemma 4.9 its eigenvalues must be $3 - 2\cos\frac{2\pi j}{n}$, $j = 1, \ldots, n$. Thus, the determinant of $L(C_n) + I$ equals

$$\prod_{j=1}^{n}\left(3 - 2\cos\frac{2\pi j}{n}\right),$$

which is the number of spanning trees of W_n. Another consequence of Theorem 4.7 is the following.

Corollary 4.10 *Let G be a tree with $V(G) = \{1, 2, \ldots, n\}$. Let i, j be distinct vertices of G and suppose that the unique path between i and j has length ℓ. Then $\det L(i, j \mid i, j) = \ell$.*

Proof By Theorem 4.7 $\det L(i, j \mid i, j)$ equals the number of spanning forests of G with two components, one of which contains i and the other contains j. Since there is a unique path between the two vertices, the only way of obtaining such a forest is to delete an edge on the unique ij-path. $\qquad\square$

Let G be a graph with $V(G) = \{1, 2, \ldots, n\}$. Let the eigenvalues of $L(G)$ be $\lambda_1 \geq \cdots \geq \lambda_{n-1} \geq \lambda_n = 0$. Recall that the sum of the principal minors of $L(G)$ of order $n - 1$ equals the sum of the products of the eigenvalues, taken $n - 1$ at a time. Note that each principal minor of $L(G)$ equals the number of spanning trees of G. Since $\lambda_n = 0$, the sum of the products of the eigenvalues, taken $n - 1$ at a time, equals $\lambda_1 \cdots \lambda_{n-1}$. Thus, we have proved the following result.

Theorem 4.11 *Let G be a graph with $V(G) = \{1, 2, \ldots, n\}$. Let the eigenvalues of $L(G)$ be $\lambda_1 \geq \cdots \geq \lambda_{n-1} \geq \lambda_n = 0$. Then the number of spanning trees of G equals $\lambda_1 \cdots \lambda_{n-1}/n$.*

Since a graph is connected if and only if it has a spanning tree, by Theorem 4.11 we get another proof of the fact that G is connected if and only if $\lambda_{n-1} > 0$.

The eigenvalues of $L(K_n)$ are n with multiplicity $n - 1$, and 0 with multiplicity 1. Therefore, K_n has $n^{n-1}/n = n^{n-2}$ spanning trees, which is Cayley's theorem.

4.4 Bounds for Laplacian Spectral Radius

Let G be a graph with $V(G) = \{1, 2, \ldots, n\}$. Let the eigenvalues of $L(G)$ be $\lambda_1 \geq \cdots \geq \lambda_{n-1} \geq \lambda_n = 0$. Also, let Δ be the maximum vertex degree.

It follows from the well-known maximal representation of the eigenvalues of a symmetric matrix (see Chap. 1) that $\lambda_1 \geq \Delta$. We now proceed to establish the stronger statement, that $\lambda_1 \geq \Delta + 1$.

Theorem 4.12 *Let G be a graph with $V(G) = \{1, 2, \ldots, n\}$ and with at least one edge. Let the eigenvalues of $L(G)$ be $\lambda_1 \geq \cdots \geq \lambda_{n-1} \geq \lambda_n = 0$. Then $\lambda_1 \geq \Delta + 1$.*

Proof We assume, without loss of generality, that d_1, the degree of vertex 1, is the maximum vertex degree. There exists a lower triangular matrix T with nonnegative diagonal entries such that $L(G) = TT'$. Then $d_1 = \ell_{11} = t_{11}^2$, and hence $t_{11} = \sqrt{d_1}$. Comparing the first column of both sides of $L(G) = TT'$, we see that $\ell_{i1} = \sqrt{d_1}t_{i1}$, $i = 1, 2, \ldots, n$. Thus, the first diagonal entry of $T'T$ equals

$$\sum_{i=1}^{n} t_{i1}^2 = \frac{1}{d_1} \sum_{i=1}^{n} \ell_{i1}^2 = \frac{1}{d_1}(d_1^2 + d_1) = d_1 + 1.$$

The largest eigenvalue of $T'T$ exceeds or equals the largest diagonal entry of $T'T$, and hence it exceeds or equals $d_1 + 1$. The proof follows in view of the fact that the eigenvalues of $L(G) = TT'$ and $T'T$ are the same.   □

We now obtain an upper bound for λ_1.

Theorem 4.13 *Let G be a graph with vertex set $V = \{1, \ldots, n\}$. Let L be the Laplacian of G with the maximum eigenvalue λ_1. Then*

$$\lambda_1 \leq \max\{d_i + d_j - c(i, j) : 1 \leq i < j \leq n, i \sim j\},$$

where $c(i, j)$ is the number of vertices that are adjacent to both i and j.

Proof We assume that G has at least one edge, since the result is trivial for an empty graph. Let x be an eigenvector of L corresponding to λ_1. Then $Lx = \lambda_1 x$. Choose i such that $x_i = \max_k x_k$. Furthermore, choose j such that $x_j = \min_k\{x_k : k \sim i\}$. The ith and the jth equations from the vector equation $Lx = \lambda_1 x$ can be expressed as

$$\lambda_1 x_i = d_i x_i - \sum_{k : k \sim i} x_k \tag{4.1}$$

and

$$\lambda_1 x_j = d_j x_j - \sum_{k:k \sim j} x_k. \tag{4.2}$$

From (7.2) and (7.3) we get

$$\lambda_1 x_i = d_i x_i - \sum_{k:k \sim i, k \sim j} x_k - \sum_{k:k \sim i, k \not\sim j} x_k \tag{4.3}$$

and

$$\lambda_1 x_j = d_j x_j - \sum_{k:k \sim j, k \sim i} x_k - \sum_{k:k \sim j, k \not\sim i} x_k. \tag{4.4}$$

Subtracting (4.4) from (7.8),

$$\begin{aligned}
\lambda_1 (x_i - x_j) &= d_i x_i - d_j x_j - \sum_{k:k \sim i, k \not\sim j} x_k + \sum_{k:k \sim j, k \not\sim i} x_k \\
&\le d_i x_i - d_j x_j - (d_i - c(i,j)) x_j + (d_j - c(i,j)) x_i \\
&= (d_i + d_j - c(i,j))(x_i - x_j). \tag{4.5}
\end{aligned}$$

If $x_j = x_i$ for all $j \sim i$, then from (7.2) we see that $\lambda_1 = 0$, which is not possible if the graph has at least one edge. Therefore, there exists j such that $i \sim j$ and $x_i > x_j$. Now from (4.5) we see that

$$\lambda_1 \le d_i + d_j - c(i,j),$$

and the result follows.   □

Corollary 4.14 *Let G be a graph with the vertex set $V = \{1, \ldots, n\}$. Let L be the Laplacian of G with maximum eigenvalue λ_1. Then*

$$\lambda_1 \le \max\{d_i + d_j : 1 \le i < j \le n, i \sim j\}.$$

4.5 Edge-Laplacian of a Tree

Let G be a graph with $V(G) = \{1, \ldots, n\}$ and $E(G) = \{e_1, \ldots, e_m\}$. Assign an orientation to each edge, and let Q be the $n \times m$ incidence matrix. The matrix $K = Q'Q$ has been termed the *edge-Laplacian of G*. The rows and the columns of K are indexed by $E(G)$. For $i \ne j$, the (i,j)-element of K is 0 if the edges e_i and e_j have no vertex in common. If they do have a common vertex then the (i,j)-element of K is -1 or 1 according as e_i and e_j follow each other, or not, respectively. The diagonal entries of K are all equal to 2. Note that $\operatorname{rank} K = \operatorname{rank} Q$, and it follows that the edge-Laplacian of a tree is nonsingular. In the remainder of this section we consider the edge-Laplacian of a tree and obtain a combinatorial description of its inverse.

Let T be a tree with the vertex set $\{1, \ldots, n\}$ and the edge set $\{e_1, \ldots, e_{n-1}\}$. Assign an orientation to each edge of T and let Q be the incidence matrix.

Lemma 4.15 *Let H be the $(n-1) \times n$ matrix defined as follows. The rows and the columns of H are indexed by the edges and the vertices of T, respectively. Set $h_{ij} = 1$ if edge e_i is directed away from vertex j, and $h_{ij} = 0$ otherwise. Then $HQ = I_{n-1}$.*

Proof Fix $i \neq j$. Suppose edge e_j joins vertices u, v and is directed from u to v. Then $q_{uj} = 1$, $q_{vj} = -1$ and $q_{kj} = 0$, $k \neq u$, $k \neq v$. Thus, the (i, j)-element of HQ equals

$$\sum_{k=1}^{n} h_{ik} q_{kj} = h_{iu} - h_{iv}.$$

Note that e_i is either directed away from both u and v or is directed towards both u and v. Therefore, $h_{iu} = h_{iv}$ and hence the (i, j)-element of HQ is zero. If $i = j$ then $h_{iu} = 1$ and $h_{iv} = 0$ and then $\sum_{k=1}^{n} h_{ik} q_{kj} = h_{iu} - h_{iv} = 1$. This completes the proof.   □

By Lemma 4.15 $HQH = H$ and therefore H is a g-inverse of Q. It is well known that the class of g-inverses of Q is given by $H + X(I - QH) + (I - HQ)Y$, where X and Y are arbitrary. Since $HQ = I$ by Lemma 4.15, the class of g-inverses of Q is given by $H + X(I - QH)$, where X is arbitrary. We now determine the X that produces the Moore–Penrose inverse of Q.

By Lemma 2.2, rank $HQ = $ rank $Q = n - 1$. Also rank $(I - QH) = n -$ rank $(QH) = 1$. Therefore, rank $X(I - QH) \leq 1$ and hence $X(I - QH) = uv'$ for some vectors u and v. Thus, we conclude that $Q^+ = H + uv'$ for some vectors u and v, which we now proceed to determine.

For $i = 1, \ldots, n - 1$, the graph $T \backslash e_i$ has two components, both trees, one of which is closer to the tail of e_i, while the other is closer to the head of e_i. We refer to these as the tail component and the head component of e_i, respectively. Let t_i be the number of vertices in the tail component of e_i. Let $t = (t_1, \ldots, t_{n-1})'$. It is clear from the definition of H that $H\mathbf{1} = t$.

Considering $Q^+ = H + uv'$ and $Q^+ \mathbf{1} = 0$, we get $H\mathbf{1} + (v'\mathbf{1})u = 0$, and hence $(v'\mathbf{1})u = -H\mathbf{1} = -t$.

Also, $I = Q^+ Q = HQ + uv'Q = I + uv'Q$, and hence $uv'Q = 0$. Since $Q^+ \neq H$, u and v are nonzero vectors. Hence, $v'Q = 0$ and $v = \alpha\mathbf{1}$ for some α. Therefore, $t = -(v'\mathbf{1})u = -\alpha\mathbf{1}'\mathbf{1}u = -\alpha nu$. It follows that

$$Q^+ = H + uv'$$
$$= H - \frac{t}{\alpha n}(\alpha\mathbf{1}')$$
$$= H - \frac{1}{n}t\mathbf{1}'.$$

Thus, we have obtained the formula for Q^+ given in the next result.

Theorem 4.16 *The rows and the columns of Q^+ are indexed by the edges and the vertices of T, respectively. The (i, j)-element of Q^+ is $-\frac{t_i}{n}$ if j is in the head component of e_i, and it equals $1 - \frac{t_i}{n}$ if j is in the tail component of e_i.*

Example 4.17   Consider the following tree:

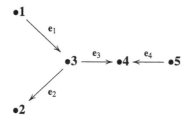

The incidence matrix is

$$Q = \begin{bmatrix} 1 & 0 & 0 & 0 \\ 0 & -1 & 0 & 0 \\ -1 & 1 & 1 & 0 \\ 0 & 0 & -1 & -1 \\ 0 & 0 & 0 & 1 \end{bmatrix}.$$

Then

$$H = \begin{bmatrix} 1 & 0 & 0 & 0 & 0 \\ 1 & 0 & 1 & 1 & 1 \\ 1 & 1 & 1 & 0 & 0 \\ 0 & 0 & 0 & 0 & 1 \end{bmatrix}$$

satisfies $HQ = I$, while the Moore–Penrose inverse of Q is given by

$$Q^+ = \frac{1}{5} \begin{bmatrix} 4 & -1 & -1 & -1 & -1 \\ 1 & -4 & 1 & 1 & 1 \\ 2 & 2 & 2 & -3 & -3 \\ -1 & -1 & -1 & -1 & 4 \end{bmatrix}.$$

It is well known (see Exercise 6) that for any matrix A, $(AA')^+ = (A^+)'A^+$. Thus, using Theorem 4.16 we may obtain an expression for L^+, where $L = QQ'$ is the Laplacian of the tree. We state the expression and omit the easy verification. We first introduce some notation. If i is a vertex and e_j an edge of T, then $f(e_j, i)$ will denote the number of vertices in the component of $T\backslash\{e_j\}$ that does not contain i. Also, for vertices i, j and edge e_k, $\alpha(i, j, e_k)$ will be -1 or 1 according as e_k is on the (i, j)-path or otherwise, respectively.

Theorem 4.18 *For $i, j = 1, \ldots, n$, the (i, j)-element of L^+ is given by*

$$\frac{1}{n^2} \sum_{k=1}^{n-1} \alpha(i, j, e_k) f(e_k, i) f(e_k, j).$$

As observed earlier, since $\text{rank } K = \text{rank } Q'Q = \text{rank } Q = n - 1$, then K is nonsingular. It is easily seen by the Cauchy–Binet formula that $\det K = n$. Since $K^{-1} = K^+ = Q^+(Q^+)'$, we may obtain an expression for K^{-1} using Theorem 4.16. Again, we only state the expression. Extending our earlier notation, let us denote by $f(e_j, e_i)$ the number of vertices in the component of $T \backslash \{e_j\}$ that does not contain e_i.

Suppose edge e_i has head u and tail v, while edge e_j has head w and tail x. We say that e_i and e_j are similarly oriented if the path joining u and w contains precisely one of x or v. Otherwise, we say that e_i and e_j are oppositely oriented.

Theorem 4.19 *For $i, j = 1, \ldots, n - 1$, the (i, j)-element of K^{-1} is given by*

$$\pm \frac{1}{n}(n - f(e_i, e_j))(n - f(e_j, e_i)),$$

where the sign is positive or negative according as e_i and e_j are similarly oriented or oppositely oriented, respectively.

Exercises

1. Let G be a graph with n vertices and let L be the Laplacian of G. Show that the number of spanning trees of G is given by $\frac{1}{n^2} \det(L + J)$.
2. Let $G \times H$ be the Cartesian product of graphs G and H. Determine $L(G \times H)$ in terms of $L(G)$ and $L(H)$. Hence, determine the Laplacian eigenvalues of $G \times H$ in terms of those of G and H.
3. Let G be a graph with n vertices and m edges. Show that $\kappa(G)$, the number of spanning trees of G, satisfies

$$\kappa(G) \le \frac{1}{n}\left(\frac{2m}{n-1}\right)^{n-1}.$$

4. Let G be a graph with vertex set $V(G) = \{1, \ldots, n\}$. Let L be the Laplacian of G with maximum eigenvalue λ_1. Show that $\lambda_1 \le n$.
5. Let T be a tree with vertices $\{1, \ldots, n\}$ and edges $\{e_1, \ldots, e_{n-1}\}$. Show that the edges of T can be oriented in such a way that the edge-Laplacian K becomes an entrywise nonnegative matrix.
6. Let A be an $m \times n$ matrix. Show that $(A')^+ = (A^+)'$ and that $(AA')^+ = (A')^+A^+$.

Basic properties of the Laplacian are discussed in the books by Biggs and by Godsil and Royle quoted in Chap. 2. Other relevant references are as follows: Sect. 4.4: [GM94, AM85, Das03], Sect. 4.5: [Bap97, Mer89, Moo95].

References and Further Reading

[AM85]   Anderson, W.N., Morley, T.D.: Eigenvalues of the Laplacian of a graph. Linear and Mul-
         tilinear Algebra **18**(2), 141–145 (1985)
[Bap97] Bapat, R.B.: Moore–Penrose inverse of the incidence matrix of a tree. Linear and Multi-
         linear Algebra **42**, 159–167 (1997)
[Das03] Das, K.Ch.: An improved upper bound for Laplacian graph eigenvalues. Linear Algebra
         Appl. **368**, 269–278 (2003)
[GM94]   Grone, R., Merris, R.: The Laplacian spectrum of a graph II. SIAM J. Discret. Math. **7**(2),
         221–229 (1994)
[Mer89] Merris, R.: An edge version of the matrix-tree theorem and the Wiener index. Linear and
         Multilinear Algebra **25**, 291–296 (1989)
[Moo95] Moon, J.W.: On the adjoint of a matrix associated with trees. Linear and Multilinear
         Algebra **39**, 191–194 (1995)

Chapter 5
Cycles and Cuts

Let G be a graph with $V(G) = \{1, \ldots, n\}$ and $E(G) = \{e_1, \ldots, e_m\}$. Assign an orientation to each edge of G and let Q be the incidence matrix. The null space of Q is called the *cycle subspace* of G whereas the row space of Q is called the *cut subspace* of G. These definitions are justified as follows.

Consider a cycle $\mathscr{C}$ in G and choose an orientation of the cycle. Let x be the $m \times 1$ incidence vector of the cycle. We claim that $Qx = 0$, that is, x is in the null space of Q. The ith element of Qx is $(Qx)_i = \sum_{j=1}^{m} q_{ij}x_j$. If vertex i and $\mathscr{C}$ are disjoint, then clearly $(Qx)_i = 0$. Otherwise there must be precisely two edges of $\mathscr{C}$ which are incident with i. Suppose e_p with endpoints i, k and e_s with endpoints i, ℓ are in $\mathscr{C}$. If e_p has head i and tail k and if e_s has head i and tail ℓ, then we have $q_{ip} = 1$, $q_{is} = 1$ and $q_{ij}x_j = 0$ for $j \neq p, j \neq s$. Also, $x_p = -x_s$. It follows that $(Qx)_i = 0$. The cases when e_p and e_s have other orientations are similar. Therefore, $(Qx)_i = 0$ for each i and hence x is in the null space of Q.

We now turn to cuts. Let $V(G) = V_1 \cup V_2$ be a partition of $V(G)$ into nonempty disjoint subsets V_1 and V_2. The set of edges with one endpoint in V_1 and the other endpoint in V_2 is called a *cut*. Denote this cut by $\mathscr{K}$. Given a cut $\mathscr{K}$ we define its incidence vector y as follows. The order of y is $m \times 1$ and its components are indexed by $E(G)$. If e_i is not in $\mathscr{K}$, then $y_i = 0$. If $e_i \in \mathscr{K}$, then $y_i = 1$ or -1 according as e_i is directed from V_1 to V_2, or from V_2 to V_1, respectively.

Let u be a vector of order $n \times 1$ defined as follows. The components of u are indexed by $V(G)$. Set $u_i = 1$ or -1 according as $i \in V_1$ or $i \in V_2$, respectively. Observe that $y' = \frac{1}{2}u'Q$ and hence y is in the row space of Q.

5.1 Fundamental Cycles and Fundamental Cuts

We continue to use the notation introduced earlier. If G is a graph with k connected components, then by Theorem 2.3 rank $Q = n - k$. Hence the dimension of the cycle subspace of G is $m - n + k$, whereas the dimension of the cut subspace of G is $n - k$. We now describe a procedure to obtain bases for these two subspaces.

© Springer-Verlag London 2014
R.B. Bapat, *Graphs and Matrices*, Universitext,
DOI 10.1007/978-1-4471-6569-9_5

The cycle subspace of G is the direct sum of the cycle subspaces of each of its connected components. A similar remark applies to the cut subspace of G. Therefore, for the purpose of determining bases for the cycle subspace and the cut subspace, we may restrict our attention to connected graphs.

Let G be a connected graph and let T be a spanning tree of G. The edges $E(G)\backslash E(T)$ are said to constitute a *cotree* of G, which we denote by T^c, the complement of T. If $e_i \in E(T^c)$ then $E(T) \cup \{e_i\}$ contains a unique cycle, which we denote by $\mathscr{C}_i$. The cycle $\mathscr{C}_i$ is called a *fundamental cycle*. The orientation of $\mathscr{C}_i$ is taken to be consistent with the orientation of e_i.

Theorem 5.1 *Let G be a connected graph with n vertices, m edges, and let T be a spanning tree of G. For each $e_i \in E(T^c)$, let x^i be the incidence vector of the fundamental cycle $\mathscr{C}_i$. Then $\{x^i : e_i \in E(T^c)\}$ forms a basis for the cycle subspace of G.*

Proof As observed earlier, x^i is in the cycle subspace of G. Note that $|E(T^c)| = m - n + 1$. Since the dimension of the cycle subspace of G is $m - n + 1$, we only need to prove that $\{x^i : e_i \in E(T^c)\}$ are linearly independent.

If $e_i \in E(T^c)$ then the fundamental cycle $\mathscr{C}_i$ contains precisely one edge, namely e_i, from $E(T^c)$, while all the remaining edges of $\mathscr{C}_i$ come from $E(T)$. Thus, e_i does not belong to any other fundamental cycle. In other words, x^i has a nonzero entry at a position where each x^j, $j \neq i$, has a zero. Hence, $\{x^i : e_i \in E(T^c)\}$ is a linearly independent set. $\square$

The procedure for finding a basis for the cut subspace of G also uses the spanning tree. Let $e_i \in E(T)$. The graph obtained by removing e_i from T is a forest with two components. Let V_1 and V_2 be the vertex sets of the two components. Then $V(G) = V_1 \cup V_2$ is a partition. We assume that e_i is directed from V_1 to V_2. Let $\mathscr{K}_i$ denote the cut of G corresponding to the partition $V_1 \cup V_2$ and let y^i be its incidence vector. The cut $\mathscr{K}_i$ is called a *fundamental cut*.

Theorem 5.2 *Let G be a connected graph with n vertices, m edges, and let T be a spanning tree of G. For each $e_i \in E(T)$, let y^i be the incidence vector of the fundamental cut $\mathscr{K}_i$. Then $\{y^i : e_i \in E(T)\}$ forms a basis for the cut subspace of G.*

Proof Since $|E(T)| = n - 1$, which is the dimension of the cut subspace of G, we only need to prove that $\{y^i : e_i \in E(T)\}$ is a linearly independent set. As in the proof of Theorem 5.1, each fundamental cut contains precisely one edge from $E(T)$ and that edge is in no other fundamental cut. Hence, $\{y^i : e_i \in E(T)\}$ is a linearly independent set. This completes the proof. $\square$

Example 5.3 Consider the graph G:

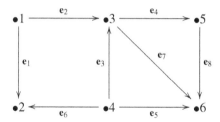

Let T be the spanning tree formed by $\{e_1, e_2, e_3, e_4, e_5\}$. The fundamental cycle associated with e_6 is $1 - 2 - 4 - 3 - 1$ and its incidence vector is

$$\begin{bmatrix} -1 & 1 & -1 & 0 & 0 & 1 & 0 & 0 \end{bmatrix}.$$

The fundamental cut associated with e_3 corresponds to the partition $V_1 = \{4, 6\}$, $V_2 = \{1, 2, 3, 5\}$ and its incidence vector is

$$\begin{bmatrix} 0 & 0 & 1 & 0 & 0 & 1 & -1 & -1 \end{bmatrix}.$$

5.2 Fundamental Matrices

Let G be a connected graph with $V(G) = \{1, \ldots, n\}$ and $E(G) = \{e_1, \ldots, e_m\}$. Let T be a spanning tree of G. We assume that $E(T) = \{e_1, \ldots, e_{n-1}\}$. Then the cotree T^c has edge set $E(T^c) = \{e_n, \ldots, e_m\}$. As usual, we assume that the edges of G have been assigned an orientation.

The fundamental cut matrix B of G is an $(n-1) \times m$ matrix defined as follows. The rows of B are indexed by $E(T)$, while the columns are indexed by $E(G)$. The ith row of B is the incidence vector of the fundamental cut $\mathscr{K}_i$ associated with $e_i, i = 1, \ldots, n-1$. Since e_i is the only edge of T that is in $\mathscr{K}_i, i = 1, \ldots, n-1$, B must be of the form $[I, B_f]$ where B_f is of order $(n-1) \times (m-n+1)$.

The fundamental cycle matrix C of G is an $(m-n+1) \times m$ matrix defined as follows. The rows of C are indexed by $E(T^c)$, while the columns are indexed by $E(G)$. The ith row of C is the incidence vector of the fundamental cycle $\mathscr{C}_i$ associated with $e_i, i = n, \ldots, m$. Since e_i is the only edge of T^c that is in $\mathscr{C}_i, i = n, \ldots, m$, C must be of the form $[C_f, I]$ where C_f is of order $(m-n+1) \times (n-1)$.

Lemma 5.4 $B_f = -C'_f$.

Proof Let Q be the incidence matrix of G. As seen earlier, each row vector of B is in the row space of Q. Also, the transpose of any row vector of C is in the null space of Q. It follows that $BC' = 0$. Therefore,

$$[I \ B_f] \begin{bmatrix} C'_f \\ I \end{bmatrix} = 0,$$

and hence $C'_f + B_f = 0$. Thus, $B_f = -C'_f$.   □

Example 5.5 Consider the graph G and the spanning tree T as in Example 5.3. Then

$$B = \begin{bmatrix} 1 & 0 & 0 & 0 & 0 & 1 & 0 & 0 \\ 0 & 1 & 0 & 0 & 0 & -1 & 0 & 0 \\ 0 & 0 & 1 & 0 & 0 & 1 & -1 & -1 \\ 0 & 0 & 0 & 1 & 0 & 0 & 0 & -1 \\ 0 & 0 & 0 & 0 & 1 & 0 & 1 & 1 \end{bmatrix}$$

and

$$C = \begin{bmatrix} -1 & 1 & -1 & 0 & 0 & 1 & 0 & 0 \\ 0 & 0 & 1 & 0 & -1 & 0 & 1 & 0 \\ 0 & 0 & 1 & 1 & -1 & 0 & 0 & 1 \end{bmatrix}.$$

Let Q be the incidence matrix of G. There is a close relationship between Q, B and C, as we see next.

Theorem 5.6 *Let Q_1 be the reduced incidence matrix obtained by deleting the last row of Q and suppose Q_1 is partitioned as $Q_1 = [Q_{11}, Q_{12}]$, where Q_{11} is of order $(n-1) \times (n-1)$. Then $B_f = Q_{11}^{-1} Q_{12}$ and $C_f = -Q'_{12}(Q'_{11})^{-1}$.*

Proof The rank of Q_{11} equals $n-1$, which is the rank of Q. Therefore, the rows of Q_1 form a basis for the row space of Q. Since each row of B is in the row space of Q, there exists a matrix Z such that $B = ZQ_1$. In partitioned form, this equation reads

$$[I \ B_f] = Z[Q_{11} \ Q_{12}].$$

It follows that $ZQ_{11} = I$ and $ZQ_{12} = B_f$. Thus, $Z = Q_{11}^{-1}$ and $B_f = Q_{11}^{-1}Q_{12}$. The second part follows, since by Lemma 5.4, $C_f = -B'_f$.   □

5.3 Minors

We continue to use the notation introduced earlier. We first consider minors of B and C containing all the rows.

Theorem 5.7 *Let G be a connected graph with n vertices, m edges, and let B be the fundamental cut matrix of G with respect to the spanning tree T. Then the following assertions hold:*

   (i) *a set of columns of B is a linearly independent set if and only if the corresponding edges of G induce an acyclic graph;*

(ii) *a set of $n - 1$ columns of B is a linearly independent set if and only if the corresponding edges form a spanning tree of G;*

(iii) *if X is a submatrix of B of order $(n - 1) \times (n - 1)$, then det X is either 0 or ± 1;*

(iv) *det BB' equals the number of spanning trees of G.*

Proof Recall that the columns of B are indexed by $E(G)$. Let Q be the incidence matrix of G. Let Q_1 be the reduced incidence matrix and let $Q_1 = [Q_{11}, Q_{12}]$ as in Theorem 5.6. By Theorem 5.6, $B = Q_{11}^{-1} Q_1$. Let Y be the submatrix of B formed by the columns $j_1, \ldots, j_k$, and let R be the submatrix of Q_1 formed by the columns with the same indices. Then $Y = Q_{11}^{-1} R$, and hence, rank Y = rank R. In particular, the columns of Y are linearly independent if and only if the corresponding columns of R are linearly independent. By Lemma 2.5, the columns of R are linearly independent if and only if the corresponding edges of G form an acyclic graph. This proves (i). Assertion (ii) follows easily from (i).

To prove (iii), note that det X is det Q_{11}^{-1} multiplied by the determinant of a submatrix of Q_1 of order $(n - 1) \times (n - 1)$. Since Q is totally unimodular (see Lemma 2.6), it follows that det X is either 0 or ± 1.

To prove (iv), first observe that, by the Cauchy–Binet formula, det $BB' = \sum (\det Z)^2$, where the summation is over all $(n - 1) \times (n - 1)$ submatrices Z of B. By (ii), det Z is nonzero if and only if the corresponding edges form a spanning tree of G, and then by (iii), det Z must be ± 1. Hence, det BB' equals the number of spanning trees of G. □

We now turn to the fundamental cycle matrix. Let C be the fundamental cycle matrix of G with respect to the spanning tree T. Recall that the columns of C are indexed by $E(G)$.

Lemma 5.8 *Columns $j_1, \ldots, j_k$ of C are linearly dependent if the subgraph of G induced by the corresponding edges contains a cut.*

Proof As usual, let Q be the incidence matrix of G. Suppose that the edges of G indexed by $j_1, \ldots, j_k$ contain a cut. Let u be the incidence vector of the cut. As observed earlier, u' is in the row space of Q and hence $u' = z'Q$ for some vector z. Then $Cu = CQ'z = 0$, since $CQ' = 0$. Note that only the coordinates of u indexed by $j_1, \ldots, j_k$ can possibly be nonzero. Thus, from $Cu = 0$ we conclude that the columns $j_1, \ldots, j_k$ are linearly dependent. □

If $E(G) = E_1 \cup E_2$ is a partition of the edge set of the connected graph G into disjoint subsets, and if E_1 does not contain a cut, then E_2 must induce a connected, spanning subgraph. We will use this observation.

Lemma 5.9 *Let X be a submatrix of C of order $(m - n + 1) \times (m - n + 1)$. Then X is nonsingular if and only if the edges corresponding to the column indices of X form a cotree of G.*

Proof Let the columns of X be indexed by $F \subset E(G)$. If X is nonsingular, then by Lemma 5.8, the subgraph induced by F does not contain a cut. Then F^c induces a connected, spanning subgraph. Since $|F^c| = n - 1$, the subgraph must be a spanning tree of G. Therefore, the edges in F form a cotree.

Conversely, suppose the edges in F form a cotree S^c, where S is a spanning tree of G. Let D be the fundamental cycle matrix with respect to S. Note that the columns of C, as well as D, are indexed by $E(G)$, listed in the same order. Since the rows of C, as well as D, are linearly independent, and since their row spaces are the same, there exists a nonsingular matrix Z of order $(m - n + 1) \times (m - n + 1)$ such that $C = ZD$. Therefore, an $(m - n + 1) \times (m - n + 1)$ submatrix of C is nonsingular if and only if the corresponding submatrix of D is nonsingular. The submatrix of D indexed by F is the identity matrix. Hence, the submatrix of C indexed by F is nonsingular. □

We now prove the converse of Lemma 5.8.

Lemma 5.10 *Let $F \subset E(G)$ and suppose the columns of C indexed by F are linearly dependent. Then the subgraph of G induced by F contains a cut.*

Proof If the subgraph of G induced by F does not contain a cut, then the subgraph of G induced by F^c is spanning and connected. Therefore the subgraph induced by F^c contains a spanning tree S of G. By Lemma 5.9, the columns of C indexed by the edges in the cotree S^c are linearly independent. These columns include all the columns indexed by F. Then the columns of F must also be linearly independent. This is a contradiction and the result is proved. □

Our next objective is to show that the fundamental cut matrix and the fundamental cycle matrix are totally unimodular.

Lemma 5.11 *Let G be a connected graph with n vertices, m edges, and let B be the fundamental cut matrix of G with respect to the spanning tree T. Then B is totally unimodular.*

Proof Consider a $k \times k$ submatrix D of B, and suppose D is indexed by $E_1 \subset E(T)$ and $E_2 \subset E(G)$. If $k = n - 1$, then by Theorem 5.7, det D is either 0 or ± 1. So, suppose $k < n - 1$. If det $D = 0$ then there is nothing to prove. So, suppose D is nonsingular. Then the columns of B indexed by E_2 are linearly independent, and by Theorem 5.7, the corresponding edges induce an acyclic subgraph of G. We may extend this subgraph to a spanning tree S, using only edges from T. The submatrix of B formed by the columns corresponding to the edges in S is a matrix of order $(n - 1) \times (n - 1)$, and it is nonsingular by Theorem 5.7. Thus, det $S = \pm 1$. We may expand det S using columns coming from the identity matrix and therefore det $S = \pm$ det D. Hence, det $D = \pm 1$. □

Lemma 5.12 *Let G be a connected graph with n vertices, m edges, and let C be the fundamental cycle matrix of G with respect to the spanning tree T. Then C is totally unimodular.*

Proof Recall that if B is the fundamental cut matrix with respect to the spanning tree T, then $B = [I, B_f]$ and $C = [-B'_f, I]$. Consider a submatrix F of C. If F is a submatrix of $-B'_f$, then it follows by Lemma 5.11 that det F is either 0 or ± 1. If F contains some part from the identity matrix, then we may expand det F along the columns coming from the identity matrix and again conclude that det F is either 0 or ± 1. $\qquad\square$

We saw in Theorem 5.7 that det BB' equals the number of spanning trees in G. We now give an interpretation of the principal minors of BB'.

Theorem 5.13 *Let G be a connected graph with n vertices, m edges, and let B be the fundamental cut matrix of G with respect to the spanning tree T. Let $E \subset E(T)$ and let $BB'[E|E]$ be the submatrix of BB' with rows and columns indexed by E. Then* det $BB'[E|E]$ *equals the number of ways of extending E^c to a spanning tree of G.*

Proof Let $|E| = k$. By the Cauchy–Binet formula,

$$\det BB'[E|E] = \sum_{F \subset E(G), |F|=k} (\det B[E|F])^2, \qquad (5.1)$$

where $B[E|F]$ denotes the submatrix of B indexed by the rows in E and the columns in F. Note that since $B[E(T)|E(T)]$ is the identity matrix, $B[E|F]$ is nonsingular if and only if $B[E(T)|F \cup E^c]$ is nonsingular, in which case by Lemma 5.11, det $B[E|F] = \pm 1$. Now $B[E(T)|F \cup E^c]$ is nonsingular if and only if the edges $F \cup E^c$ form a spanning tree of G, and hence the result follows by (5.1). $\qquad\square$

Corollary 5.14 *Let G be a connected graph with n vertices, m edges, and let B be the fundamental cut matrix of G with respect to the spanning tree T. Let $e_i \in E(T)$ and let $BB'(e_i|e_i)$ be the submatrix of BB' with row and column indexed by e_i deleted. Then* det $BB'(e_i|e_i)$ *equals the number of spanning forests of G with two components, such that the endpoints of e_i are in different components.*

Proof By Theorem 5.13, det $BB'(e_i|e_i)$ equals the number of ways of extending e_i to a spanning tree of G, which is precisely the number as asserted in the result. $\quad\square$

It may be mentioned that the theory of fundamental matrices may be developed for undirected graphs, resulting in 0–1 matrices. The treatment is similar, except the underlying field is that of integers modulo 2.

Exercises

1. Let G be a connected graph with n vertices, m edges, B the fundamental cut matrix, and C the fundamental cycle matrix of G. Show that the $m \times m$ matrix $\begin{bmatrix} B \\ C \end{bmatrix}$ is nonsingular.

2. Let $\mathcal{K}_i$ be a cut in G with incidence vector $x^i, i = 1, \ldots, n-1$. Suppose $x^1, \ldots, x^{n-1}$ are linearly independent. Show that all nonzero $(n-1) \times (n-1)$ minors of the matrix $X = [x^1, \ldots, x^{n-1}]$ are equal.

3. Let G be a connected graph with n vertices, m edges, and let C be the fundamental cycle matrix of G with respect to the spanning tree T. Let $E \subset E(T)^c$. Show that det $CC'[E|E]$ equals the number of ways of extending E^c to a cotree of G.

4. Let G be a connected graph with n vertices, m edges, and let B be the fundamental cut matrix of G with respect to the spanning tree T. Let T_1 be a subtree of T. Show that det $BB'[E(T_1)|E(T_1)]$ equals det $L(V(T_1)|V(T_1))$, where L is the Laplacian matrix of G.

5. Let G be a connected planar graph and let G^* be its dual. Let T be a spanning tree of G and let T^* be its dual spanning tree of G^*. Show that the fundamental cut matrix of G with respect to T equals the fundamental cycle matrix of G^* with respect to T^*.

The material in this chapter is generally covered in most basic texts, but the level and the depth to which it is covered may vary. We list below only two selected references: [D74] is recommended for an elementary treatment, while [R89], Chap. 1, is more advanced. The statements and the proofs of several results in Sect. 5.3 have not appeared in the literature in the present form.

References and Further Reading

[D74] Deo, N.: Graph Theory with Applications to Engineering and Computer Science. Prentice-Hall Inc, New Jersey (1974)
[R89] Recski, A.: Matroid Theory and Its Applications in Electric Network Theory and in Statics. Algorithms and Combinatorics, vol. 6. Springer, Berlin (1989)

Chapter 6
Regular Graphs

A graph is said to be *regular* if all its vertices have the same degree. If the degree of each vertex of G is k, then G is said to be k-regular. Examples of regular graphs include cycles, complete graphs and complete bipartite graphs with bipartite sets of the same cardinality.

6.1 Perron–Frobenius Theory

We prove those aspects of the Perron–Frobenius theorem that are required for application to graphs. First we introduce some notation.

For a vector x we write $x \geq 0$ to indicate that each coordinate of x is nonnegative, while $x > 0$ means that each coordinate of x is positive. Similar notation applies to matrices. For matrices A and B, $A \geq B$ denotes that $A - B \geq 0$. Similarly, $A > B$ denotes that $A - B > 0$. The spectral radius $\rho(A)$ of a square matrix A is the maximum modulus of an eigenvalue of A. The spectral radius of a graph G, denoted $\rho(G)$, is the spectral radius of the adjacency matrix of the graph.

Lemma 6.1 *Let G be a connected graph with n vertices, and let A be the adjacency matrix of G. Then $(I + A)^{n-1} > 0$.*

Proof Clearly, $(I + A)^{n-1} \geq I + A + A^2 + \cdots + A^{n-1}$. Since G is connected, for any $i \neq j$, there is an (ij)-path, and the length of the path can be at most $n - 1$. Thus, the (i, j)-element of $I + A + A^2 + \cdots + A^{n-1}$ is positive. If $i = j$, then clearly, the (i, j)-element of $I + A + A^2 + \cdots + A^{n-1}$ is positive. Therefore, $(I + A)^{n-1} > 0$ and the proof is complete. $\qquad\square$

Lemma 6.2 *Let G be a connected graph with n vertices, and let A be the adjacency matrix of G. If $x \geq 0$ is an eigenvector of A, then $x > 0$.*

Proof If $Ax = \mu x$, then clearly, $\mu > 0$. We have $(I + A)^{n-1}x = (1 + \mu)^{n-1}x$. By Lemma 6.1, $(I + A)^{n-1} > 0$ and it follows that $x > 0$. $\qquad\square$

© Springer-Verlag London 2014
R.B. Bapat, *Graphs and Matrices*, Universitext,
DOI 10.1007/978-1-4471-6569-9_6

Theorem 6.3 *Let G be a connected graph with $n \geq 2$ vertices, and let A be the adjacency matrix of G. Then the following assertions hold:*

(i) *A has an eigenvalue $\lambda > 0$ and an associated eigenvector $x > 0$.*
(ii) *for any eigenvalue $\mu \neq \lambda$ of A, $-\lambda \leq \mu < \lambda$. Furthermore, $-\lambda$ is an eigenvalue of A if and only if G is bipartite.*
(iii) *if u is an eigenvector of A for the eigenvalue λ, then $u = \alpha x$ for some α.*

Proof Let

$$P^n = \left\{ y \in \mathbb{R}^n : y_i \geq 0, \quad i = 1, \ldots, n; \quad \sum_{i=1}^n y_i = 1 \right\}.$$

We define $f : P^n \to P^n$ as $f(y) = \frac{1}{\sum_i (Ay)_i} Ay$, $y \in P^n$. Since G is connected A has no zero column and hence for any $y \in P^n$, Ay has at least one positive coordinate. Hence, f is well-defined. Clearly, P^n is a compact, convex set, and f is a continuous function from P^n to itself. By the well-known Brouwer's fixed point theorem, there exists $x \in P^n$ such that $f(x) = x$. If we let $\lambda = \sum_{i=1}^n (Ax)_i$, then it follows that $Ax = \lambda x$. Now $(1 + \lambda)^{n-1} x = (I + A)^{n-1} x > 0$ by Lemma 6.1. Hence, $(1 + \lambda)^{n-1} x > 0$ and therefore $x > 0$. This proves (i).

Let $\mu \neq \lambda$ be an eigenvalue of A and let z be an associated eigenvector, so that $Az = \mu z$. Then

$$|\mu||z_i| \leq \sum_{j=1}^n a_{ij}|z_j|, \quad i = 1, \ldots, n. \tag{6.1}$$

Using the vector x in (i), we get from (6.1),

$$|\mu| \sum_{i=1}^n x_i |z_i| \leq \sum_{i=1}^n x_i \sum_{j=1}^n a_{ij}|z_j|$$

$$= \sum_{j=1}^n |z_j| \sum_{i=1}^n a_{ij} x_i$$

$$= \lambda \sum_{j=1}^n x_j |z_j|. \tag{6.2}$$

It follows from (6.2) that $|\mu| \leq \lambda$, that is, $-\lambda \leq \mu < \lambda$. If $\mu = -\lambda$ is an eigenvalue of A with the associated eigenvector z, then we see from the above proof that equality must hold in (6.1) for $i = 1, \ldots, n$; that is,

$$\lambda |z_i| = \sum_{j=1}^n a_{ij}|z_j| = \sum_{j \sim i} |z_j|. \tag{6.3}$$

Thus, $|z| = (|z_1|, \ldots, |z_n|)'$ is an eigenvector of A for λ, and, as seen in the proof of (i), $|z_i| > 0$, $i = 1, \ldots, n$. Also, $Az = -\lambda z$ gives

$$-\lambda z_i = \sum_{j \sim i} z_j, \, i = 1, \ldots, n. \tag{6.4}$$

From (6.3) and (6.4),

$$\lambda |z_i| = \left| \sum_{j \sim i} z_j \right| \leq \sum_{j \sim i} |z_j| \leq \lambda |z_i|.$$

Therefore, for any i, z_j has the same sign for all $j \sim i$.

Let $V_1 = \{i \in V(G) : z_i > 0\}$ and $V_2 = \{i \in V(G) : z_i < 0\}$. Then it can be seen that G is bipartite with the bipartition $V(G) = V_1 \cup V_2$. If G is bipartite, then by Theorem 3.14, $-\lambda$ is an eigenvalue of A. This completes the proof of (ii).

Let u be an eigenvector of A for the eigenvalue λ. We may choose a scalar β such that $x - \beta u \geq 0$ and $x - \beta u$ has a zero coordinate. If $x - \beta u \neq 0$, then it is an eigenvector of A for the eigenvalue λ with all the coordinates nonnegative. As seen in the proof of (i), we may conclude that all its coordinates must be positive, a contradiction. Therefore, $x - \beta u = 0$ and, setting $\alpha = 1/\beta$, (iii) is proved. □

The eigenvalue λ of G, as in (i) of Theorem 6.3, is called the *Perron eigenvalue* of G, and the associated eigenvector x is called a Perron eigenvector. Note that by (ii) of the theorem, the Perron eigenvalue of G is the same as the spectral radius $\rho(G)$. The Perron eigenvector is unique, up to a scalar multiple, as seen in (iii) of the theorem. For graphs that are not necessarily connected we may prove the following.

Theorem 6.4 *Let G be a graph with n vertices, and let A be the adjacency matrix of G. Then $\rho(G)$ is an eigenvalue of G and there is an associated nonnegative eigenvector.*

Proof Let $G_1, \ldots, G_p$ be the connected components of G, and let $A_1, \ldots, A_p$ be the corresponding adjacency matrices. We assume, without loss of generality, that $\rho(G_1) = \max_i \rho(G_i)$. Then by Theorem 6.3 there is a vector $x > 0$ such that $A_1 x = \rho(G_1)x$. The vector obtained by augmenting x by zeros is easily seen to be an eigenvector of A corresponding to the eigenvalue $\rho(G) = \rho(G_1)$. □

In view of Theorem 6.4, we refer to $\rho(G)$ as the Perron eigenvalue of the graph G, which may be connected or otherwise. We now turn to some monotonicity properties of the Perron root.

Lemma 6.5 *Let G be a connected graph with n vertices, and let $H \neq G$ be a spanning, connected subgraph of G. Then $\rho(G) > \rho(H)$.*

Proof Let A and B be the adjacency matrices of G and H, respectively. By Theorem 6.3 there exist vectors $x > 0$, $y > 0$, such that $Ax = \rho(G)x$, $By = \rho(H)y$. Since $0 \neq A - B \geq 0$ and since $x > 0$, $y > 0$, then $y'Ax > y'Bx$. But $y'Ax = y'(\rho(G)x) = \rho(G)y'x$ and $y'Bx = \rho(H)y'x$. Therefore, $\rho(G) > \rho(H)$. □

Lemma 6.6 *Let G be a connected graph and let A be the adjacency matrix of G. Let $\mu > 0$, $x \geq 0$ be such that $Ax \geq \mu x$, $Ax \neq \mu x$. Then $\mu < \rho(G)$.*

Proof By Theorem 6.3, there exists $y > 0$ such that $Ay = \rho(G)y$. We have

$$\begin{aligned}
(\rho(G) - \mu)y'x &= y'\rho(G)x - \mu y'x \\
&= y'Ax - \mu y'x \\
&= y'(Ax - \mu x) > 0,
\end{aligned}$$

since $Ax - \mu x \geq 0$, $Ax - \mu x \neq 0$. As $y'x > 0$, it follows that $\mu < \rho(G)$. □

Lemma 6.7 *Let G be a connected graph with n vertices and let $H \neq G$ be a vertex-induced subgraph of G. Then $\rho(G) > \rho(H)$.*

Proof Let A and B be the adjacency matrices of G and H, respectively. Then B is a principal submatrix of A. We assume, without loss of generality, that $A = \begin{pmatrix} B & C \\ C' & E \end{pmatrix}$. By Theorem 6.4, there exists $z \geq 0$ such that $Bz = \rho(H)z$. Let $x = [z', 0]'$. Then

$$\begin{aligned}
Ax &= \begin{pmatrix} B & C \\ C' & E \end{pmatrix} \begin{pmatrix} z \\ 0 \end{pmatrix} \\
&= \begin{pmatrix} Bz \\ C'z \end{pmatrix} \\
&= \begin{pmatrix} \rho(H)z \\ C'z \end{pmatrix} \\
&\geq \rho(H)x.
\end{aligned}$$

If $Ax = \rho(H)x$, then by Lemma 6.2, $x > 0$, which is a contradiction. Thus $Ax - \rho(H)x \geq 0$, $Ax - \rho(H)x \neq 0$. It follows from Lemma 6.6 that $\rho(G) > \rho(H)$. □

Theorem 6.8 *Let G be a connected graph and let $H \neq G$ be a subgraph of G. Then $\rho(G) > \rho(H)$.*

Proof Note that H must have a connected component H_1 such that $\rho(H) = \rho(H_1)$, and H_1 is a spanning subgraph of a vertex-induced, connected subgraph H_2 of G. If $H_2 = G$, then by Lemma 6.5, $\rho(H_1) < \rho(H_2)$. If $H_2 \neq G$, then by Lemma 6.7, $\rho(H_2) < \rho(G)$. Also by Lemma 6.5, $\rho(H_1) \leq \rho(H_2)$ (equality holds if $H_1 = H_2$) and hence $\rho(H_1) < \rho(G)$. This completes the proof. □

If G is a connected graph, then by Theorem 6.3 (iii), $\rho(G)$ is an eigenvalue of G with geometric multiplicity 1. Since the adjacency matrix is symmetric, the following result is immediate.

Theorem 6.9 *Let G be a connected graph with n vertices. Then $\rho(G)$ is an eigenvalue of G with algebraic multiplicity 1.*

The following result for regular graphs is a consequence of the results obtained thus far.

Theorem 6.10 *Let G be a k-regular graph. Then $\rho(G)$ equals k, and it is an eigenvalue of G. It has algebraic multiplicity 1 if G is connected.*

Proof Let A be the adjacency matrix of G. By Theorem 6.3 there exists $0 \neq x \geq 0$ such that $Ax = \rho(G)x$. Since G is k-regular, $A\mathbf{1} = k\mathbf{1}$. Hence, $\mathbf{1}'Ax = k(\mathbf{1}'x)$, and also $\mathbf{1}'Ax = \rho(G)(\mathbf{1}'x)$. Therefore, $\rho(G) = k$. If G is connected then by Theorem 6.9 k has algebraic multiplicity 1. $\qquad\square$

We now obtain some bounds for the Perron eigenvalue.

Theorem 6.11 *Let G be a connected graph with n vertices, and let A be the adjacency matrix of G. Then for any $y, z \in \mathbb{R}^n$, $y \neq 0$, $z > 0$,*

$$\frac{y'Ay}{y'y} \leq \rho(G) \leq \max_i \left\{ \frac{(Az)_i}{z_i} \right\}. \tag{6.5}$$

Equality holds in the first inequality if and only if y is an eigenvector of A corresponding to $\rho(G)$. Similarly, equality holds in the second inequality if and only if z is an eigenvector of A corresponding to $\rho(G)$.

Proof The first inequality follows from the extremal representation for the largest eigenvalue of a symmetric matrix. The assertion about equality also follows from the general result about symmetric matrices.

To prove the second inequality, suppose that for $z > 0$, $\rho(G) > \max_i \left\{ \frac{(Az)_i}{z_i} \right\}$, $i = 1, \ldots, n$. Then $Az < \rho(G)z$. Let $x > 0$ be the Perron vector of A so that $Ax = \rho(G)x$. It follows that $\rho(G)z'x = z'Ax = x'Az < \rho(G)x'z$, which is a contradiction. The assertion about equality is easily proved. $\qquad\square$

Corollary 6.12 *Let G be a connected graph with n vertices and m edges. Let $d_1 \geq \cdots \geq d_n$ be the vertex degrees. Then the following assertions hold:*

(i) $\frac{2m}{n} \leq \rho(G) \leq d_1$;

(ii) $\dfrac{1}{2m} \displaystyle\sum_{i=1}^{n} \sum_{i<j, j\sim i} \sqrt{d_i d_j} \leq \rho(G) \leq \max_i \left\{ \dfrac{1}{d_i} \sum_{j\sim i} \sqrt{d_i d_j} \right\}$.

Furthermore, equality holds in any of the above inequalities if and only if G is regular.

Proof To prove (i), set $y = z = 1$, whereas to prove (ii), set $y = z = [\sqrt{d_1}, \ldots, \sqrt{d_n}]'$ in Theorem 6.11.   □

We conclude this section with an application of the Perron–Frobenius theorem to obtaining a proof of Turán's theorem.

Theorem 6.13 *Let G be a graph with n vertices, m edges, and no triangles. Then* $m \le \frac{n^2}{4}$.

Proof Let A be the adjacency matrix of G. Let $\rho(G) = \lambda_1 \ge \lambda_2 \ge \cdots \ge \lambda_n$ be the eigenvalues of A. Let, if possible, $m > \frac{n^2}{4}$. By (i), Corollary 6.12,

$$\lambda_1 \ge \frac{2m}{n} > \sqrt{m}. \tag{6.6}$$

Recall that the trace of A^2 equals $\sum_{i=1}^{n} \lambda_i^2$, and it also equals $2m$. It follows from (6.6) that $2m = \sum_{i=1}^{n} \lambda_i^2 > m + \sum_{i=2}^{n} \lambda_i^2$, and hence

$$\lambda_1^2 > m > \sum_{i=2}^{n} \lambda_i^2. \tag{6.7}$$

By the Perron–Frobenius theorem, $\lambda_1 \ge |\lambda_i|$, $i = 2, \ldots, n$, and hence

$$\left| \sum_{i=2}^{n} \lambda_i^3 \right| \le \sum_{i=2}^{n} |\lambda_i|^3 \le \lambda_1 \left(\sum_{i=2}^{n} |\lambda_i|^2 \right) < \lambda_1^3, \tag{6.8}$$

in view of (6.7).

Each triangle in a graph gives rise to 6 closed walks of length 3. Thus, the number of triangles in G equals $\frac{1}{6}$ trace $A^3 = \frac{1}{6} \sum_{i=1}^{n} \lambda_i^3$. Now

$$\frac{1}{6} \sum_{i=1}^{6} \lambda_i^3 = \frac{\lambda_1^3}{6} + \frac{\sum_{i=2}^{n} \lambda_i^3}{6},$$

which must be positive by (6.8). This is a contradiction, as G has no triangles, and hence $m \le \frac{n^2}{4}$.   □

6.2 Adjacency Algebra of a Regular Graph

If B is an $n \times n$ matrix, then the algebra generated by B is defined as the set of all linear combinations of $I, B, B^2, \ldots$. In other words, the algebra generated by B is the set of matrices that are polynomials in B. If G is a graph with adjacency matrix A, then the algebra generated by A is called the *adjacency algebra* of G. The following result due to Hoffman characterizes regular graphs in terms of the adjacency algebra. Recall that the matrix of all 1 s is denoted by J.

Theorem 6.14 *Let G be a graph with n vertices. Then G is a connected, regular graph if and only if J is in the adjacency algebra of G.*

Proof Let A be the adjacency matrix of G. First suppose that J is in the adjacency algebra of G. Then there exist real numbers $\alpha_0, \alpha_1, \ldots, \alpha_t$ for some t such that

$$J = \alpha_0 I + \alpha_1 A + \cdots + \alpha_t A^t. \tag{6.9}$$

It follows from (6.9) that $AJ = JA$. Note that if $d_1, \ldots, d_n$ are the vertex degrees, then

$$AJ = \begin{bmatrix} d_1 \\ \vdots \\ d_n \end{bmatrix} \mathbf{1}',$$

while $JA = \mathbf{1}[d_1, \ldots, d_n]$. Therefore, $AJ = JA$ implies that $d_i = d_j$ for all i, j, and hence G is regular. If G is disconnected, then there exist vertices i, j such that there is no (ij)-walk. Then the (i, j)-entry of A^p is 0, $p \geq 0$, and clearly this contradicts (6.9). Hence, G is connected.

Conversely, suppose G is connected and k-regular. Let $p(\lambda)$ be the minimal polynomial of A. Since k is an eigenvalue of A, then $p(\lambda) = (\lambda - k)q(\lambda)$ for some polynomial $q(\cdot)$. From $p(A) = 0$ we get $Aq(A) = kq(A)$. Thus, each column of $q(A)$ is an eigenvector of A corresponding to $k = \rho(G)$. By Theorem 6.3 each column of $q(A)$ must be a multiple of $\mathbf{1}$. Since $q(A)$ is symmetric it follows that $q(A) = \alpha J$ for some α. Thus, J is in the adjacency algebra of G. $\square$

The constant α in the proof of Theorem 6.14 can be determined explicitly. Let $k = \lambda_1 > \lambda_2 > \cdots > \lambda_p$ be the distinct eigenvalues of A. Then $p(\lambda) = (\lambda - k)(\lambda - \lambda_2) \cdots (\lambda - \lambda_p) = (\lambda - k)q(\lambda)$ is the minimal polynomial of A. As seen in the proof of Theorem 6.14, $q(A) = \alpha J$ for some α. The eigenvalues of $q(A)$ are $q(k)$, and $q(\lambda_2) = \cdots = q(\lambda_p) = 0$. Comparing the largest eigenvalue of $q(A)$ and αJ we see that $q(k) = \alpha n$, and hence $\alpha = \frac{q(k)}{n}$.

6.3 Complement and Line Graph of a Regular Graph

If G is a regular graph then there are simple relations between its adjacency matrix and Laplacian matrix, as well as the corresponding matrices of G^c, the complement of G, and G_ℓ, the line graph of G. These relations lead to several statements about the characteristic polynomials of regular graphs, some of which will be proved now.

Theorem 6.15 *Let G be a k-regular graph with n vertices. Let A and $\overline{A}$ be the adjacency matrices of G and G^c, respectively. If $k = \lambda_1, \lambda_2, \ldots, \lambda_n$ are the eigenvalues of A, then $n - 1 - \lambda_1, -1 - \lambda_2, \ldots, -1 - \lambda_n$ are the eigenvalues of $\overline{A}$.*

Proof Since G is k-regular, $\mathbf{1}$ is an eigenvector of A corresponding to k. Set $z = \frac{1}{\sqrt{n}}\mathbf{1}$, and let P be an orthogonal matrix with its first column equal to z such that $P'AP = \text{diag}(\lambda_1, \lambda_2, \ldots, \lambda_n)$. Since $A + \overline{A} = J - I$, it follows that

$$P'\overline{A}P = P'(J - I - A)P$$
$$= P'JP - I - P'AP$$
$$= \text{diag}(n - 1 - \lambda_1, -1 - \lambda_2, \ldots, -1 - \lambda_n),$$

where we have used the fact that any column of P other than the first column is orthogonal to the first column. Hence, the eigenvalues of $\overline{A}$ are as asserted. □

Let G be a graph with adjacency matrix A. The characteristic polynomial of A is given by $\det(\lambda I - A)$. We refer to this polynomial as the characteristic polynomial of G and denote it $\phi(G, \lambda)$.

Corollary 6.16 *Let G be a k-regular graph with n vertices. Then*

$$\phi(G^c, \lambda) = (-1)^n \frac{\lambda + k + 1 - n}{\lambda + k + 1} \phi(G, -\lambda - 1).$$

Proof Let $k = \lambda_1, \lambda_2, \ldots, \lambda_n$ be the eigenvalues of G. Then $\phi(G, \lambda) = (\lambda - \lambda_1)(\lambda - \lambda_2) \cdots (\lambda - \lambda_n)$. By Theorem 6.15, $n - 1 - \lambda_1, -1 - \lambda_2, \ldots, -1 - \lambda_n$ are the eigenvalues of G^c, and hence

$$\phi(G^c, \lambda) = (\lambda - n + 1 + \lambda_1)(\lambda + 1 + \lambda_2) \cdots (\lambda + 1 + \lambda_n).$$

Therefore,

$$\frac{\phi(G^c, \lambda)}{\phi(G, -\lambda - 1)} = \frac{(\lambda - n + 1 + \lambda_1)(\lambda + 1 + \lambda_2) \cdots (\lambda + 1 + \lambda_n)}{(-\lambda - 1 - \lambda_1)(-\lambda - 1 - \lambda_2) \cdots (-\lambda - 1 - \lambda_n)}$$
$$= (-1)^n \frac{\lambda - n + 1 + \lambda_1}{\lambda + 1 + \lambda_1}$$

and the proof is complete. □

Theorem 6.17 *Let G be a k-regular graph with n vertices. Then the number of spanning trees of G is given by $\frac{1}{n}\phi'(G, \lambda)|_{\lambda=k}$.*

Proof If A and L are the adjacency matrix and the Laplacian matrix of G, respectively, then $L = kI - A$. Let $k, \lambda_2, \ldots, \lambda_n$ be the eigenvalues of A. Then the eigenvalues of L are $0, k - \lambda_2, \ldots, k - \lambda_n$. By Theorem 4.11 the number of spanning trees of G is given by $\frac{1}{n}(k - \lambda_2) \cdots (k - \lambda_n)$. Since $\phi(G, \lambda) = (\lambda - k)(\lambda - \lambda_2) \cdots (\lambda - \lambda_n)$, we see that $\frac{1}{n}\phi'(G, \lambda)|_{\lambda=k} = \frac{1}{n}(k - \lambda_2) \cdots (k - \lambda_n)$ and the proof is complete. □

We now turn to line graphs. Let G be a graph with $V(G) = \{1, \ldots, n\}$ and $E(G) = \{e_1, \ldots, e_m\}$. Recall that the line graph G_ℓ of G has vertex set $E(G)$. For $i \neq j$, e_i and e_j are said to be adjacent if they have a common vertex. If G is k-regular then G_ℓ is $(2k - 2)$-regular. We first prove a preliminary result. Recall the definition of the $0 - 1$ incidence matrix M of G given in Chap. 2.

Lemma 6.18 *Let G be a graph with n vertices. Let A and B be the adjacency matrices of G and of G_ℓ, respectively. If M is the incidence matrix of G, then $M'M = B + 2I$. Furthermore, if G is k-regular then $MM' = A + kI$.*

Proof Any diagonal entry of $M'M$ clearly equals 2. If e_i and e_j are edges of G then the (i, j)-element of $M'M$ is 1 if e_i and e_j have a common vertex, and 0 otherwise. Hence, $M'M = B + 2I$. To prove the second part, note that for a k-regular graph, $MM' = -L + 2kI$, where L is the Laplacian of G. Hence, $A = kI - L = MM' - kI$. Therefore, $MM' = A + kI$. $\qquad\square$

We note in passing a consequence of Lemma 6.18.

Corollary 6.19 *Let G be a graph. If μ is an eigenvalue of G_ℓ then $\mu \geq -2$.*

Proof Let B be the adjacency matrix of G_ℓ. By Lemma 6.18 $B + 2I = M'M$ is positive semidefinite. If μ is an eigenvalue of B then $\mu + 2$, being an eigenvalue of a positive semidefinite matrix, must be nonnegative. Hence, $\mu \geq -2$. $\qquad\square$

Theorem 6.20 *Let G be a k-regular graph with n vertices. Then*

$$\phi(G_\ell, \lambda) = (\lambda + 2)^{\frac{n}{2}(k-2)}\phi(G, \lambda + 2 - k).$$

Proof Let A and B be the adjacency matrices of G and of G_ℓ, respectively. Let M be the incidence matrix of G. If G has m edges then M is of order $n \times m$. Let $k = \lambda_1, \lambda_2, \ldots, \lambda_n$ be the eigenvalues of A. By Lemma 6.18 the eigenvalues of MM' are $2k, \lambda_2 + k, \ldots, \lambda_n + k$. Note that the eigenvalues of $M'M$ are given by the eigenvalues of MM', together with 0 with multiplicity $m - n$. Therefore, again by Lemma 6.18, the eigenvalues of B are $2k - 2, \lambda_2 + k - 2, \ldots, \lambda_n + k - 2$, and -2 with multiplicity $m - n$. Since

$$\phi(G, \lambda) = (\lambda - k)(\lambda - \lambda_2) \cdots (\lambda - \lambda_n),$$

then

$$\phi(G, \lambda + 2 - k) = (\lambda + 2 - 2k)(\lambda + 2 - k - \lambda_2) \cdots (\lambda + 2 - k - \lambda_n).$$

Also,

$$\phi(G_\ell, \lambda) = (\lambda + 2 - 2k)(\lambda + 2 - k - \lambda_2) \cdots (\lambda + 2 - k - \lambda_n)(\lambda + 2)^{n-m}.$$

Hence,

$$\frac{\phi(G_\ell, \lambda)}{\phi(G, \lambda + 2 - k)} = (\lambda + 2)^{m-n}. \tag{6.10}$$

Since G is k-regular then $2m = nk$ and hence $m - n = \frac{n}{2}(k - 2)$. Substituting in (6.10) the result is proved. □

6.4 Strongly Regular Graphs and Friendship Theorem

Let G be a k-regular graph with n vertices. The graph G is said to be *strongly regular* with parameters (n, k, a, c) if the following conditions hold:

  (i) G is neither complete, nor empty;
 (ii) any two adjacent vertices of G have a common neighbours;
(iii) any two nonadjacent vertices of G have c common neighbours.

For example, C_5 is strongly regular with parameters $(5, 2, 0, 1)$, while the Petersen graph is strongly regular with parameters $(10, 3, 0, 1)$.

Lemma 6.21 *Let G be a strongly regular graph with parameters (n, k, a, c) and let A be the adjacency matrix of G. Then*

$$A^2 = kI + aA + c(J - I - A). \tag{6.11}$$

Proof Let $B = kI + aA + c(J - I - A)$. Any diagonal entry of A^2 clearly equals k and so does any diagonal entry of B. If i and j are adjacent vertices of G, then the (i, j)-element of B is a. The (i, j)-element of A^2 equals the number of walks of length 2 from i to j, which also equals a since G is strongly regular. A similar argument shows that the (i, j)-elements of A^2 and B are equal when i and j are nonadjacent. Hence, $A^2 = B$ and the proof is complete. □

The following statement, which is essentially a converse of Lemma 6.21, is easy to prove using the definition of a strongly regular graph.

Lemma 6.22 *Let G be a graph which is neither complete nor empty, and let A be the adjacency matrix of G. Then G is strongly regular if A^2 is a linear combination of A, I and J.*

We now determine the eigenvalues of a strongly regular graph.

Theorem 6.23 *Let G be a strongly regular graph with parameters (n, k, a, c) and let A be the adjacency matrix of G. Let $\Delta = (a - c)^2 + 4(k - c)$. Then any eigenvalue of A is either k or $\frac{1}{2}(a - c \pm \sqrt{\Delta})$.*

Proof Since G is k-regular, k is an eigenvalue of A with $\mathbf{1}$ as the corresponding eigenvector. Let $\mu \neq k$ be an eigenvalue of A with y as the corresponding eigenvector, so that $Ay = \mu y$. Note that $y'\mathbf{1} = 0$. By Lemma 6.21,

$$A^2 = kI + aA + c(J - I - A),$$

and hence

$$A^2 y = ky + aAy + c(-y - Ay). \tag{6.12}$$

It follows from (6.12) that

$$\mu^2 = k + a\mu + c(-1 - \mu).$$

Thus, μ is a solution of the equation

$$x^2 - (a - c)x - (k - c) = 0.$$

The solutions of this equation are $\frac{1}{2}(a - c \pm \sqrt{\Delta})$, which must be the possible values of μ. □

Theorem 6.24 *Let G be a connected, strongly regular graph with parameters* (n, k, a, c). *Let* $\Delta = (a - c)^2 + 4(k - c)$. *Then the numbers*

$$m_1 = \frac{1}{2}\left(n - 1 + \frac{(n-1)(c-a) - 2k}{\sqrt{\Delta}}\right)$$

and

$$m_2 = \frac{1}{2}\left(n - 1 - \frac{(n-1)(c-a) - 2k}{\sqrt{\Delta}}\right)$$

are nonnegative integers.

Proof By Theorem 6.23 the eigenvalues of G are k and $\frac{1}{2}(a - c \pm \sqrt{\Delta})$. Since G is connected, k has multiplicity 1. Let m_1 and m_2 be the multiplicities of the remaining two eigenvalues. Then

$$1 + m_1 + m_2 = n. \tag{6.13}$$

Since the sum of the eigenvalues equals the trace of the adjacency matrix, which is 0, we have

$$k + \frac{m_1}{2}(a - c + \sqrt{\Delta}) + \frac{m_2}{2}(a - c - \sqrt{\Delta}) = 0. \tag{6.14}$$

From (6.13) $m_2 = n - 1 - m_1$. Substituting in (6.14) we get

$$k + \frac{m_1}{2}(a - c + \sqrt{\Delta}) + \frac{n - 1 - m_1}{2}(a - c - \sqrt{\Delta}) = 0.$$

Thus,

$$m_1\sqrt{\Delta} + k + \frac{n - 1}{2}(a - c - \sqrt{\Delta}) = 0,$$

or

$$m_1 = \frac{1}{\sqrt{\Delta}}\left(-k - \frac{n-1}{2}(a - c - \sqrt{\Delta})\right) = \frac{1}{2}\left(n - 1 + \frac{(n-1)(c-a) - 2k}{\sqrt{\Delta}}\right),$$

as asserted. The value of m_2 is obtained using $m_2 = n - 1 - m_1$ and is seen to be

$$m_2 = \frac{1}{2}\left(n - 1 - \frac{(n-1)(c-a) - 2k}{\sqrt{\Delta}}\right).$$

Since m_1 and m_2 are multiplicities, they must be nonnegative integers and the proof is complete.   □

We recall the result (see Corollary 3.3) that if G is a connected graph then the diameter of G is less than the number of distinct eigenvalues of G.

Theorem 6.25 *Let G be a connected regular graph with exactly three distinct eigenvalues. Then G is strongly regular.*

Proof Let G have n vertices and suppose it is k-regular. Since G has three distinct eigenvalues, by the preceding remark, it has diameter at most 2. Since G is connected and is neither complete nor empty, its diameter cannot be 0 or 1 and hence it must be 2. Since G is k-regular one of its eigenvalues must be k. Let the other two eigenvalues be θ and τ, and let $p(x) = (x - \theta)(x - \tau)$. Then $(A - kI)p(A) = 0$. Since G is connected k has multiplicity 1, and hence the null space of $A - kI$ is spanned by $\mathbf{1}$. As $(A - kI)p(A) = 0$, each column of $p(A)$ is a multiple of $\mathbf{1}$. Furthermore, since $p(A)$ is symmetric it follows that $p(A) = \alpha J$ for some α. Thus,

$$(A - \theta I)(A - \tau I) = \alpha J.$$

It follows that A^2 is a linear combination of A, I and J. We conclude by Lemma 6.22, that G is strongly regular.   □

As an application of the integrality condition obtained in Theorem 6.24, we prove the next result, known as the *friendship theorem*.

Theorem 6.26 *Let G be a graph in which any two distinct vertices have exactly one common neighbour. Then G has a vertex that is adjacent to every other vertex, and, more precisely, G consists of a number of triangles with a common vertex.*

Proof First observe that from the given hypotheses it easily follows that G is connected.

Let i and j be nonadjacent vertices of G, and let $N(i)$ and $N(j)$ be their respective neighbour sets. With $x \in N(i)$, we associate the $y \in N(j)$, which is the unique common neighbour of x and j. Set $y = f(x)$ and observe that f is a one-to-one mapping from $N(i)$ to $N(j)$. Indeed, if $z \in N(i)$, $z \neq x$, satisfies $f(z) = y$, then x and z would have two common neighbours, namely i and y, which is a contradiction.

Therefore, f is one-to-one and hence $|N(i)| \leq |N(j)|$. We may similarly show that $|N(j)| \leq |N(i)|$ and hence $|N(i)| = |N(j)|$.

Suppose G is k-regular. By the hypotheses, G must be strongly regular with parameters $(n, k, 1, 1)$. By Theorem 6.24, $m_1 - m_2 = \frac{k}{\sqrt{k-1}}$ is an integer. So $k - 1$ divides k^2, which is possible only if $k = 0$ or 2. If $k = 0$ then, since G is connected, $n = 1$. Then the theorem holds vacuously. If $k = 2$ then, in view of the hypothesis that any two vertices have exactly one common neighbour, G must be the complete graph on 3 vertices and again the theorem holds.

Finally, suppose G is not regular. Then by the first part of the proof there must be adjacent vertices i and j with unequal degrees. Let x be the unique common neighbour of i and j, and we assume, without loss of generality, that the degrees of i and x are unequal. Let y be any vertex other than i, j and x. If y is not adjacent to both i and j, then then degrees of i and j would be equal to that of y, which is not possible. Hence, y is adjacent to either i or j. Similarly y is adjacent to either i or x. Since y cannot be adjacent to both j and x (j and x already have i as their common neighbour) then y must be adjacent to i. It follows that all the vertices other than x and j are adjacent to i. The proof also shows that G consists of a number of triangles with i as the common vertex.   □

According to Theorem 6.26, if any two individuals in a group have exactly one common friend, then there must be a person who is a friend of everybody. This justifies the name "friendship theorem." The following figure shows an example of a graph satisfying the hypotheses of Theorem 6.26.

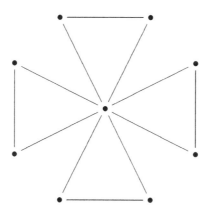

6.5 Graphs with Maximum Energy

Let G be a graph with $V(G) = \{1, \ldots, n\}$. Let A be the adjacency matrix of G, and let $\lambda_1, \ldots, \lambda_n$ be the eigenvalues of A. Recall that the energy of G is defined as $\varepsilon(G) = \sum_{i=1}^{n} |\lambda_i|$. We now obtain some bounds for the energy of a graph and consider the cases of equality.

Theorem 6.27 *Let G be a graph with n vertices, m edges, and suppose $2m \geq n$. Then*

$$\varepsilon(G) \leq \frac{2m}{n} + \sqrt{(n-1)\left[2m - \left(\frac{2m}{n}\right)^2\right]}. \tag{6.15}$$

Proof As noted before, trace $A^2 = \sum_{i=1}^{n} \lambda_i^2 = 2m$. Hence,

$$\sum_{i=2}^{n} \lambda_i^2 = 2m - \lambda_1^2. \tag{6.16}$$

It follows from (6.16) and the Cauchy–Schwarz inequality that

$$\sum_{i=2}^{n} |\lambda_i| \leq \sqrt{(n-1)(2m - \lambda_1^2)}. \tag{6.17}$$

From (6.17) we thus have

$$\varepsilon(G) \leq \lambda_1 + \sqrt{(n-1)(2m - \lambda_1^2)}. \tag{6.18}$$

Consider the function

$$f(x) = x + \sqrt{(n-1)(2m - x^2)}.$$

It is easily seen that $f(x)$ decreases on the interval $\sqrt{\frac{2m}{n}} < x \leq \sqrt{2m}$. By (i), Corollary 6.12, $\lambda_1 \geq \frac{2m}{n}$, and hence

$$\sqrt{\frac{2m}{n}} \leq \frac{2m}{n} \leq \lambda_1.$$

Hence, $f(\lambda_1) \leq f(\sqrt{\frac{2m}{n}})$. This fact and (6.18) immediately give (6.15) and the proof is complete. □

We now consider the case of equality in (6.15). The eigenvalues of K_n are $n-1$ (with multiplicity 1) and -1 (with multiplicity $n-1$). Hence, it can be seen that equality holds in (6.15) for K_n. If n is even, then equality holds in (6.15) for the graph consisting of $\frac{n}{2}$ copies of K_2 as well.

Conversely, suppose equality holds in (6.15). From the proof of Theorem 6.27, we see that $\lambda_1 = \frac{2m}{n}$. Thus, by Corollary 6.12, G is k-regular with $k = \frac{2m}{n}$. Furthermore, equality must hold in the Cauchy–Schwarz inequality used in the proof of Theorem 6.27, and hence

$$|\lambda_i| = \frac{\sqrt{2m - (2m/n)^2}}{\sqrt{n-1}}, \quad i = 2, \ldots, n.$$

Now there are three possibilities:

(i) G has two eigenvalues with equal absolute value: The eigenvalues must be of the same multiplicity as the sum of the eigenvalues is 0. Then the eigenvalues are symmetric with respect to 0, and hence by Theorem 3.14, G must be bipartite. Also the diameter of G is 1, and hence G must be a disjoint union of edges.

(ii) G has two eigenvalues with distinct absolute values: Again the diameter of each component of G is 1, and hence each component of G is a complete graph. Since G is k-regular with $k = \frac{2m}{n}$, it follows that G must be K_n.

(iii) G has three eigenvalues with distinct absolute values equal to $\frac{2m}{n}$ or $\frac{\sqrt{2m-(2m/n)^2}}{\sqrt{n-1}}$: In this case it follows by Theorem 6.25 that G is strongly regular.

Theorem 6.28 *Let G be a graph with n vertices, m edges, and suppose $2m \leq n$. Then*

$$\varepsilon(G) \leq 2m. \tag{6.19}$$

Proof Since $2m$ is the sum of the vertex degrees and $2m \leq n$, G must have $n - 2m$ isolated vertices. Let H be the graph obtained from G by removing the $n - 2m$ isolated vertices. Then H has $2m$ vertices and m edges. By Theorem 6.27 $\varepsilon(G) = \varepsilon(H) \leq 2m$, and the proof is complete. □

By the discussion of the case of equality in Theorem 6.27 it follows that equality holds in (6.19) if and only if G is a disjoint union of isolated vertices and edges. In the next result we give a bound on the energy, without assuming any hypothesis on the number of vertices and edges.

Theorem 6.29 *Let G be a graph with n vertices. Then*

$$\varepsilon(G) \leq \frac{n}{2}(1 + \sqrt{n}). \tag{6.20}$$

Proof Let G have m edges. First suppose $2m \geq n$. Let

$$f(x) = \frac{2x}{n} + \sqrt{(n-1)\left(2x - \left(\frac{2x}{n}\right)^2\right)}, \quad \frac{n}{2} \leq x \leq \frac{n^2}{2}.$$

We claim that the maximum of $f(x)$ over x in the interval $[\frac{n}{2}, \frac{n^2}{2}]$ is attained at $x = \frac{n^2 + n\sqrt{n}}{4}$. We sketch the proof of this claim:

(i) A tedious calculation shows that $f'(x) = 0$ has two roots, $x = \frac{n^2 + n\sqrt{n}}{4}$ and $x = \frac{n^2 - n\sqrt{n}}{4}$.

(ii) when $x = \frac{n^2 + n\sqrt{n}}{4}$, $f(x) = \frac{n}{2}(1 + \sqrt{n})$.

(iii) when $x = \frac{n^2 - n\sqrt{n}}{4}$, $f(x) = \frac{n}{2}(1 + \sqrt{n}) - \sqrt{n}$.

(iv) at $x = \frac{n}{2}$ and at $x = \frac{n^2}{2}, f(x) = n$.

Examining the value of f at the critical points and the boundary points of the interval $[\frac{n}{2}, \frac{n^2}{2}]$, we conclude that $f(x)$ attains its maximum at $x = \frac{n^2 + n\sqrt{n}}{4}$, and the claim is proved. Substituting this value of x in place of m in (6.15), we see that (6.20) is proved.

If $2m \leq n$, by Theorem 6.28 $\varepsilon(G) \leq n$, and (6.20) follows immediately. $\square$

As before, we conclude that equality holds in (6.20) if and only if G is strongly regular, in which case, the parameters can be seen to be (n, k, a, c), where

$$k = \frac{n + \sqrt{n}}{2}, \quad a = c = \frac{n + 2\sqrt{n}}{4}.$$

The existence of an infinite family of such graphs is known. However, we do not venture into the vast literature on the existence and construction of strongly regular graphs.

Theorem 6.29 provides an upper bound on the energy of a graph with n vertices. The bound is attained for some values of n for which the existence of certain strongly regular graphs, as described above, can be ascertained. For other values of n the problem of finding a graph with maximum energy among all graphs with n vertices remains open.

Exercises

1. Let G be a connected graph. Let μ be an eigenvalue of G with an associated nonnegative eigenvector. Show that $\mu = \rho(G)$.
2. Let G be a graph with $\rho(G) < 2$. Show that G must be acyclic and the degree of any vertex is at most 3.
3. The join $G_1 + G_2$ of graphs G_1 and G_2 is defined as $G_1 + G_2 = (G_1^c \cup G_2^c)^c$. If G_i is a k_i-regular graph with n_i vertices, $i = 1, 2$, show that

$$\frac{\phi(G_1 + G_2, \lambda)}{\phi(G_1, \lambda)\phi(G_2, \lambda)} = \frac{\lambda^2 - (k_1 + k_2)\lambda + k_1 k_2 - n_1 n_2}{(\lambda - k_1)(\lambda - k_2)}.$$

4. If G is a strongly regular graph then show that G^c is strongly regular. Determine the parameters of G^c in terms of those of G.
5. If G is k-regular then show that

$$\varepsilon(G) \leq k + \sqrt{k(n-1)(n-k)}.$$

Conclude that if G is a 3-regular graph with n vertices then $\varepsilon(G) \leq \varepsilon(K_n)$.
6. Let G be a graph with n vertices and let A be the adjacency matrix of A. Suppose A is nonsingular. Show that $\varepsilon(G) \geq n$.

We mention [BR97, BP94] as references for Perron–Frobenius theory. Sections 6.2–6.4 follow the treatment in [Cam78]. For more on strongly regular graphs, see [God93]. Section 6.5 is based on [KM01].

References and Further Reading

[BR97]   Bapat, R.B., Raghavan, T.E.S.: Encyclopedia of Mathematics and Its Applications. Nonnegative Matrices and Applications. Cambridge University Press, Cambridge (1997)
[BP94]   Berman, A., Plemmons, R.J.: Nonnegative Matrices in the Mathematical Sciences, Classics in Applied Mathematics, 9. SIAM, Philadelphia (1994)
[Cam78]   Cameron, P., J.: Strongly regular graphs, In Selected Topics in Graph Theory L.W. Beineke and R.J. Wilson, Ed. Academic Press, New York, pp. 337–360 (1978).
[God93]   Godsil, C.D.: Algebraic Combinatorics. Chapman and Hall Inc, New York (1993)
[KM01]   Koolen, J.H., Moulton, V.: Maximal energy graphs. Adv. Appl. Math. **26**, 47–52 (2001)

Chapter 7
Line Graph of a Tree

Let G be a graph with $V(G) = \{1, \ldots, n\}$ and let $A(G)$ (or simply, A) denote the adjacency matrix of G. If Δ is the diagonal matrix of vertex degrees, then recall that $L = \Delta - A$ is the Laplacian matrix of G. The matrix $K = \Delta + A$ is called the *signless Laplacian* of G. Let Q be the vertex-edge incidence matrix of the graph obtained by orienting each edge of G, and let M be the $0 - 1$ vertex-edge incidence matrix of G. Then $L = QQ'$ and $K = MM'$.

By the eigenvalues of G we mean the eigenvalues of A. Similarly G is singular (nonsingular) if A is singular (nonsingular). The eigenvalues of L or M, will be termed as the Laplacian eigenvalues or the signless Laplacian eigenvalues of G, respectively.

The line graph of the graph G is denoted $\mathscr{L}_G$. Note that $A(\mathscr{L}_G) + 2I = M'M$. This simple fact allows us to relate the eigenvalues of $\mathscr{L}_G$, to the eigenvalues of MM', and hence to the signless Laplacian eigenvalues of G. Since $L \equiv K \bmod 2$, we can use the Matrix-Tree Theorem to derive certain statements regarding K. The central, although not the most general, result in this Chapter is (see Theorem 7.12) that the nullity of the line graph of a tree is at most 1. We obtain extensions of this result and prove several related statements.

7.1 Block Graphs

We first recall some basic facts. A block of the graph G is a maximal connected subgraph of G that has no cut-vertex. Note that if G is connected and has no cut-vertex, then G itself is a block.

If an edge of a graph is contained in a cycle, then the edge by itself cannot be a block, since it is in a larger subgraph with no cut-vertex. An edge is a block if and only if it is a cut-edge. In particular, the blocks of a tree are precisely the edges of the tree. If a block has more than two vertices, then it is 2-connected. Alternatively, a block of G may be defined as a maximal 2-connected subgraph.

© Springer-Verlag London 2014
R.B. Bapat, *Graphs and Matrices*, Universitext,
DOI 10.1007/978-1-4471-6569-9_7

The complete graph with n vertices is denoted as usual by K_n. A *block graph* is a graph in which each block is a complete graph. Examples of a block graph include a complete graph, a tree and the line graph of a tree. A block graph whose blocks are K_2, K_2, K_3, K_4 and K_5 is shown below.

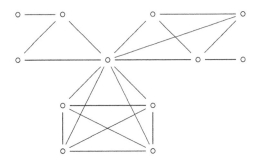

We consider the adjacency matrix of a block graph and derive a formula for its determinant.

As observed in Sect. 3.6, a tree is nonsingular if and only if it has a perfect matching. Moreover, when a tree is nonsingular, there is a formula for its inverse in terms of alternating paths. Since a tree is a block graph, it is natural to investigate the adjacency matrix of a general block graph.

A pendant vertex is a vertex of degree 1. A block is called a pendant block if it has only one cut-vertex or if it is the only block in that component. A graph is called even (odd) if it has an even (odd) number of vertices. An isolated vertex in a graph is considered to be a block of the graph.

Let G be a block graph and let $B_1, \ldots, B_k$ be the blocks of G. If $S \subset \{1, \ldots, k\}$, then G_S will denote the subgraph of G induced by the blocks $B_i, i \in S$. We first prove a preliminary result.

Lemma 7.1 *Let G be a block graph with n vertices. Let $B_1, \ldots, B_k$ be the blocks of G where B_i is the complete graph with b_i vertices, $i = 1, \ldots, k$. Let $(\alpha_1, \ldots, \alpha_k)$ be a k-tuple of nonnegative integers satisfying the following conditions:*

(i) $\displaystyle\sum_{i=1}^{k} \alpha_i = n$

(ii) *for any nonempty $S \subset \{1, \ldots, k\}$,*

$$\sum_{i \in S} \alpha_i \leq |V(G_S)|. \tag{7.1}$$

If B_i is a pendant block, then α_i equals either b_i or $b_i - 1$.

Proof Clearly by (7.1), with $S = \{i\}$, we must have $0 \leq \alpha_i \leq b_i$. Setting $S = \{1, \ldots, k\} \setminus \{i\}$ in (7.1) we see from (i) and (ii) that $\alpha_1 + \cdots + \alpha_n - \alpha_i = n - \alpha_i \leq n - b_i + 1$, and hence $\alpha_i \geq b_i - 1$. That completes the proof. $\square$

In the next result we obtain a formula for the determinant of the adjacency matrix of a block graph. The proof is based on induction and the details may be skipped in the first reading.

Theorem 7.2 *Let G be a block graph with n vertices. Let $B_1, \ldots, B_k$ be the blocks of G. Let A be the adjacency matrix of G. Then*

$$\det A = (-1)^{n-k} \sum (\alpha_1 - 1) \cdots (\alpha_k - 1) \qquad (7.2)$$

where the summation is over all k-tuples $(\alpha_1, \ldots, \alpha_k)$ of nonnegative integers satisfying the following conditions:

(i) $\displaystyle\sum_{i=1}^{k} \alpha_i = n$

(ii) *for any nonempty $S \subset \{1, \ldots, k\}$,*

$$\sum_{i \in S} \alpha_i \leq |V(G_S)|. \qquad (7.3)$$

Proof We prove the result by induction on k. The result is clearly true for a block graph with one block. Assume the result to be true for a block graph with at most $k - 1$ blocks and proceed. We consider two cases.

Case (i): The graph G has a pendant block with exactly 2 vertices.

Let $B_1 = K_2$ be a pendant block in G. Let the vertices of B_1 be 1 and 2, where 1 is pendant. We assume that the vertex 2 is in blocks $B_1, \ldots, B_p$ and we further assume that among these blocks, the first q blocks $B_1, \ldots, B_q$ are equal to K_2. It is possible that $q = 1$. Let $G_1 = G \backslash 1$. The matrix A has the form

$$A = \begin{pmatrix} 0 & 1 & 0 & \cdots & 0 \\ 1 & 0 & & \cdots & \\ 0 & & & & \\ \vdots & \vdots & & A_1 & \\ 0 & & & & \end{pmatrix}, \qquad (7.4)$$

where A_1 is the adjacency matrix of $G_2 = G_1 \backslash 2$. It follows from (7.4) that $\det A = -\det A_1$. The graph G_2 has $k - q$ blocks given by $C_i, i = q + 1, \ldots, k$, defined as follows: $C_i = B_i \backslash 2, i = q + 1, \ldots, p$ and $C_i = B_i, i = p + 1, \ldots, k$. First consider the case when $q > 1$ and that there is at least one pendant block among $B_2, \ldots, B_q$. let us assume, without loss of generality, that B_2 is a pendant block. We also assume that $V(B_2) = \{2, 3\}$. Then the first and the third columns of A are identical and $\det A = 0$. Also if $(\alpha_1, \ldots, \alpha_k)$ is a k-tuple of nonnegative integers satisfying the conditions in the Theorem, then at least one of α_1 or α_2 must equal 1 and hence the summation in (7.2) is zero. Therefore the result is proved in this case. We therefore assume that none of the blocks among $B_2, \ldots, B_q$ is pendant.

Consider

$$(-1)^{n-k} \sum (\alpha_1 - 1) \cdots (\alpha_k - 1) \tag{7.5}$$

where the summation is over all k-tuples $(\alpha_1, \ldots, \alpha_k)$ of nonnegative integers satisfying conditions (i), (ii) of the Theorem.

By Lemma 7.1, α_1 equals either 1 or 2. If $\alpha_1 = 1$, then the corresponding term in (7.5) is zero, so we assume $\alpha_1 = 2$. Then it follows from (ii) that none of $\alpha_2, \ldots, \alpha_q$ can be 2. For example, if $\alpha_2 = 2$, then $\alpha_1 + \alpha_2 = 4$, whereas the graph induced by blocks B_1 and B_2 has 3 vertices, thus violating (7.3). Thus each of $\alpha_2, \ldots, \alpha_q$ is either 0 or 1, and again we may assume that $\alpha_2 = \cdots = \alpha_q = 0$. In view of these observations (7.5) equals

$$(-1)^{n-k}(-1)^{q-1} \sum (\alpha_{q+1}-1) \cdots (\alpha_k-1) = -(-1)^{n-2-(k-q)} \sum (\alpha_{q+1}-1) \cdots (\alpha_k-1). \tag{7.6}$$

The summation in (7.6) is over all $(k - q)$-tuples $(\alpha_{q+1}, \ldots, \alpha_k)$ of nonnegative integers satisfying the following conditions:

(i) $\displaystyle\sum_{i=q+1}^{k} \alpha_i = n - 2$

(ii) for any nonempty $S \subset \{q + 1, \ldots, k\}$,

$$\sum_{i \in S} \alpha_i \leq |V(G_S)|. \tag{7.7}$$

By the induction assumption, the right side of (7.6) equals $- \det A_1$ and since $\det A = - \det A_1$, it follows that $\det A$ equals (7.5). Therefore the proof is complete in this case.

Case (ii): The graph G does not have a pendant block with exactly 2 vertices.

Let B_1 be a pendant block of G and let $V(B_1) = \{1, \ldots, b_1\}$, $b_1 \geq 3$. Let b_1 be the cut vertex in B_1. Let $H = G \backslash \{B_1 \backslash b_1\}$ be the block graph with $n - b_1 + 1$ vertices having blocks $B_2, \ldots, B_k$ and let A_1 be the adjacency matrix of H. After a suitable relabeling of the vertices in G we may write

$$A = \begin{pmatrix} D & C \\ C' & A_1 \end{pmatrix} \tag{7.8}$$

where D is the adjacency matrix of the subgraph induced by the vertex set $V(B_1 \backslash b_1)$, C is a $(b_1 - 1) \times (n - b_1 + 1)$ matrix with $c_{ij} = 1$ if $j = b_1$, and $c_{ij} = 0$ otherwise. Thus $D = J - I$ where J is the matrix of all ones and C has the form

$$\begin{pmatrix} 1 & 0 & \cdots & 0 \\ \vdots & \vdots & & \vdots \\ 1 & 0 & \cdots & 0 \end{pmatrix}.$$

Note that the first row of A_1 corresponds to the vertex b_1.

By the Schur complement formula [see (1.3)] we have

$$\det A = \det D \cdot \det(A_1 - C'D^{-1}C). \tag{7.9}$$

We note the following simple facts: (i) $\det D = (-1)^{b_1-2}(b_1 - 2)$ (ii) $D^{-1} = \frac{1}{b_1-2}J - I$ (iii) The matrix $C'D^{-1}C$ has all entries zero except the entry in the first row and the first column, which is $(-1)^{b_1}\frac{b_1-1}{\det D}$.

Let M be the adjacency matrix of $H \backslash b_1$. Thus M is the submatrix of A_1 formed by its last $n - b_1$ rows and columns. It follows from (i)–(iii) that

$$\det (A_1 - C'D^{-1}C) = \det A_1 - (-1)^{b_1}\frac{b_1 - 1}{\det D} \det M. \tag{7.10}$$

From (7.9) and (7.10) we have

$$\det A = (\det D) \left(\det A_1 - (-1)^{b_1}\frac{b_1 - 1}{\det D} \det M \right)$$
$$= (\det D)(\det A_1) - (-1)^{b_1}(b_1 - 1) \det M$$
$$= (-1)^{b_1-2}(b_1 - 2) \det A_1 - (-1)^{b_1}(b_1 - 1) \det M. \tag{7.11}$$

We assume that b_1 is in blocks B_1 and $B_2, \ldots, B_p$. If there are any blocks equal to K_2 containing b_1, then we enumerate them as $B_2, \ldots, B_q$. The remaining blocks, which do not contain b_1 are $B_{p+1}, \ldots, B_k$. If there are no blocks equal to K_2 containing b_1, then we set $q = 1$. The graph $H \backslash b_1$ has $k - q$ blocks given by $C_i, i = q+1, \ldots, k$, defined as follows: $C_i = B_i \backslash b_1, i = q + 1, \ldots, p$ and $C_i = B_i, i = p+1, \ldots, k$. Consider

$$(-1)^{n-k} \sum (\alpha_1 - 1) \cdots (\alpha_k - 1) \tag{7.12}$$

where the summation is over all k-tuples $(\alpha_1, \ldots, \alpha_k)$ of nonnegative integers satisfying (i) and (ii) of the Theorem.

By Lemma 7.1, α_1 equals either b_1 or $b_1 - 1$. If $\alpha_1 = b_1$, then each of $\alpha_2, \ldots, \alpha_q$ must be either 0 or 1. For example, if $\alpha_2 = 2$, then $\alpha_1 + \alpha_2 = b_1 + 2$, whereas the graph induced by blocks B_1 and B_2 has $b_1 + 1$ vertices, thus violating (7.3). If $\alpha_i = 1$ for some $i \in \{2, \ldots, q\}$, then the corresponding term in (7.12) is zero and hence we assume $\alpha_2 = \cdots = \alpha_q = 0$. Then it can be seen by the induction hypothesis, that the sum of the terms in (7.12) corresponding to $\alpha_1 = b_1$ equals $-(-1)^{b_1}(b_1 - 1) \det M$.

Similarly, by the induction hypothesis, the sum of the terms in (7.12) corresponding to $\alpha_1 = b_1 - 1$ equals $(-1)^{b_1-2}(b_1 - 2) \det A_1$. It follows that the sum in (7.12) equals

$$(-1)^{b_1-2}(b_1 - 2) \det A_1 - (-1)^{b_1}(b_1 - 1) \det M,$$

which is $\det A$ by (7.11). That completes the proof. □

As observed earlier in Sect. 3.6, a tree is nonsingular if and only if it has a perfect matching. We now derive this result from Theorem 7.2.

Corollary 7.3 *Let T be a tree with n vertices and let A be the adjacency matrix of T. Then A is nonsingular over reals if and only if T has a perfect matching.*

Proof First suppose that A is nonsingular. Then at least one term in the summation in (7.2) must be nonzero. If $(\alpha_1, \ldots, \alpha_{n-1})$ is an $(n-1)$-tuple of nonnegative integers satisfying (i) and (ii) of Theorem 7.2 and if the corresponding term in (7.2) is nonzero, then each α_i is either 0 or 2. Moreover if two edges have a common vertex, then the corresponding α's cannot both be nonzero, in view of (ii). Thus there must be a perfect matching in T and $\alpha_i = 2$ if and only if the corresponding edge is in the matching.

Conversely, suppose T has a perfect matching. As noted in the first part of the proof, a nonzero term in the summation in (7.2) corresponds to a perfect matching. We invoke the elementary fact, easily proved by induction, that if a tree has a perfect matching then it must be unique. Thus there must be precisely one nonzero term in the summation in (7.2) which renders det A nonzero. Thus A is nonsingular and the proof is complete. □

Example 7.4 Consider the block graph

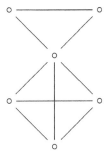

Its adjacency matrix is

$$A = \begin{pmatrix} 0 & 1 & 1 & 0 & 0 & 0 \\ 1 & 0 & 1 & 0 & 0 & 0 \\ 1 & 1 & 0 & 1 & 1 & 1 \\ 0 & 0 & 1 & 0 & 1 & 1 \\ 0 & 0 & 1 & 1 & 0 & 1 \\ 0 & 0 & 1 & 1 & 1 & 0 \end{pmatrix}.$$

We see, by Theorem 7.2, that det $A = (-1)^{6-2}((3-1)(3-1)+(2-1)(4-1)) = 7$.

A tree with no perfect matching is an example of a singular block graph. There are other examples. The following block graph is singular since the adjacency matrix has two identical columns corresponding to the pendant vertices.

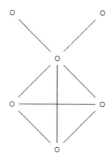

7.2 Signless Laplacian Matrix

We will repeatedly use the following fact (see Sect. 1.2): If A and B are matrices of order $m \times n$ and $n \times m$, respectively, where $m \geq n$, then the eigenvalues of AB are the same as the eigenvalues of BA, along with 0 with a (possibly further) multiplicity of $m - n$.

Lemma 7.5 *Let G be a connected graph with n vertices and let K be the signless Laplacian of G. If G is bipartite, then K is singular with rank $n - 1$. If G is non-bipartite, then K is nonsingular.*

Proof Let M be the $0 - 1$ incidence matrix of G so that $K = MM'$. By Lemma 2.17, the rank of M is $n - 1$ if G is bipartite, and n otherwise. Since rank $K = $ rank M, the result follows.   □

Corollary 7.6 *Let G be a graph with n vertices and let K be the signless Laplacian of G. Then the rank of K equals n minus the number of bipartite components of G.*

Lemma 7.7 *Let G be a connected graph with n vertices, let L be the Laplacian and let K be the signless Laplacian of G. There exists an orthogonal matrix P such that $L = PKP'$ if and only if G is bipartite.*

Proof First suppose that G is bipartite and let $V(G) = X \cup Y$ be a bipartition of G. Let P be the diagonal matrix with $p_{ii} = 1$ if $i \in X$ and $p_{ii} = -1$ if $i \in Y$. Then it can be seen that $PM = Q$, where Q is the vertex-edge incidence matrix of the directed graph obtained from G by orienting each edge from X to Y. Thus $PKP' = PMM'P' = QQ' = L$.

Conversely, let $L = PKP'$ where P is orthogonal. Since L is singular, so is K, and by Lemma 2.17, G is bipartite.   □

Let G be a graph with $V(G) = \{1, \ldots, n\}$. A subgraph of G is called acyclic if it has no cycles, or equivalently, each of its components is a tree. A subgraph of G whose components are trees, or unicyclic graphs with odd cycles, is called a TU-subgraph of G. If the TU-subgraph H of G contains c unicyclic graphs and trees $T_1, \ldots, T_k$, then the weight $w(H)$ of H is defined by

$$w(H) = 4^c \prod_{i=1}^{k} (1 + e(T_i)),$$

where $e(T_i)$ denotes the number of edges of T_i. The weight of an acyclic subgraph is defined similarly with $c = 0$.

Theorem 7.8 *Let G be a graph with $V(G) = \{1, \ldots, n\}$. Let M be the $0-1$ vertex-edge incidence matrix of G, and let $K = MM'$ be the signless Laplacian. Let W be a nonempty, proper subset of $V(G)$. Then the determinant of $K(W|W)$ equals*

$$\sum_{H} 4^{c(H)},$$

where the summation runs over all spanning TU-subgraphs H of G with $c(H)$ unicyclic components, and $|W|$ tree components, each containing a vertex of W.

Proof By the Cauchy-Binet formula, det $K(W|W)$ equals the sum of squares of the determinants of the square submatrices $M(W|F)$ where F is a set of $n - |W|$ edges of G. By Lemma 2.18, the only nonzero contributions to this sum come from substructures R of G, each component of which is a rootless tree, or a unicyclic graph with an odd cycle, and the contribution is $(\pm 2)^2 = 4$ for each unicyclic component. The rootless trees may be supplied with the roots (which are necessarily in W) and the resulting graph is a spanning TU-subgraph of G. This gives the required description of the determinant of $K(W|W)$.   □

Theorem 7.9 *Let G be a graph with $V(G) = \{1, \ldots, n\}$. Let L and K be the Laplacian and the signless Laplacian of G, respectively. Let $p_L(x) = x^n + \ell_1 x^{n-1} + \cdots + \ell_{n-1} x$ and $p_K(x) = x^n + q_1 x^{n-1} + \cdots + q_n$ be the characteristic polynomials of L and K respectively. Then*

(i) $\ell_j = (-1)^j \sum_{F_j} w(F_j), j = 1, 2, \ldots, n - 1$; *where the summation runs over all acyclic subgraphs F_j of G with j edges.*
(ii) $q_j = (-1)^j \sum_{H_j} w(H_j), j = 1, 2, \ldots, n$; *where the summation runs over all TU-subgraphs H_j of G with j edges.*

Proof (i) Note that ℓ_j equals $(-1)^j$ times the sum of the $j \times j$ principal minors of L. Thus

$$\ell_j = (-1)^j \sum_{|W|=n-j} \det L(W|W).$$

By Theorem 4.7, the determinant of $L(W|W)$ equals the number of spanning forests of G with $n - j$ components (and hence, with j edges) in which each component contains a vertex of W. Fix a spanning forest F_j of G with $n - j$ trees $T_1, \ldots, T_{n-j}$. Since T_k has $1 + e(T_k)$ vertices, $k = 1, \ldots, n - j$, such a forest will feature $\prod_{k=1}^{n-j}(1 + e(T_k)) = w(F_j)$ times in the summation above. Hence $\ell_j = (-1)^j \sum_{F_j} w(F_j)$, where the summation runs over all acyclic subgraphs F_j of G with j edges.

(ii) As in (i), q_j equals $(-1)^j$ times the sum of the $j \times j$ principal minors of K. Thus

$$q_j = (-1)^j \sum_{|W|=n-j} \det K(W|W).$$

By Theorem 7.8, the determinant of $K(W|W)$ equals the number of spanning TU-subgraphs of G with j edges in which each component is either unicyclic, or a tree containing a vertex of W. Fix a spanning TU-subgraph H_j of G. Then it contributes $w(H_j)$ to the summation above. Hence $q_j = (-1)^j \sum_{H_j} w(H_j)$, where the summation runs over all TU-subgraphs F_j of G with j edges. □

7.3 Nullity of the Line Graph of a Tree

Lemma 7.10 *Let A be an $n \times n$ integer matrix such that* $\det A = \pm 1$. *Then any rational eigenvalue of A is ± 1.*

Proof Let $f(\lambda) = a_n\lambda^n + a_{n-1}\lambda^{n-1} + \cdots a_1\lambda + a_0$ be the characteristic polynomial of A. We use the basic result that if p/q is a root of f, where p and q are coprime integers, then q divides a_n and p divides a_0. Note that $a_n = 1$ and $a_0 = \pm\det A = \pm 1$. It follows that $p/q = \pm 1$ and the proof is complete. □

Lemma 7.11 *Let T be a tree with n vertices. If the integer $\mu > 1$ is a Laplacian eigenvalue of T, then any eigenvector of the Laplacian for μ has no zero coordinate. If the integer $\mu > 1$ is a signless Laplacian eigenvalue of T, then any eigenvector of the signless Laplacian for μ has no zero coordinate.*

Proof Let the integer $\mu > 1$ be an eigenvalue of L with x as the corresponding eigenvector. Let, if possible, x have a zero coordinate, and without loss of generality we assume $x_n = 0$. Let the vertex n have degree k. We may partition L as

$$L = \begin{pmatrix} L_1 & 0 & \cdots & 0 & \cdot \\ 0 & L_2 & \cdots & 0 & \cdot \\ \vdots & \vdots & \ddots & \vdots & \cdot \\ 0 & 0 & \cdots & L_k & \cdot \\ \cdot & \cdot & \cdots & \cdot & k \end{pmatrix}.$$

Writing the equation $Lx = \mu x$ in partitioned form we observe that there exists i, $1 \leq i \leq k$, and a nonzero vector z (which is a subvector of x) such that $L_i z = \mu z$. Note that L_i equals the Laplacian of a component (which is a tree) of $T - \{n\}$, with a 1 added to one of its diagonal elements. By Laplace expansion and the Matrix-Tree Theorem we find that $\det L_i = 1$. Then by Lemma 7.10 we conclude that $\mu = \pm 1$, which is a contradiction. Therefore x has no zero coordinate.

As observed in the proof of Lemma 7.7, $L = PKP'$, where P is a diagonal matrix with ± 1 along the diagonal. Thus any eigenvector of K is obtained from an eigenvector of L by negating sum of its entries. Thus the statement about an eigenvector of the signless Laplacian follows.   □

Theorem 7.12 *The nullity of the line graph of a tree is at most 1.*

Proof Recall that $M'M = 2I + A(\mathscr{L}_T)$, and hence the nullity of $\mathscr{L}_T$ equals the multiplicity of 2 as an eigenvalue of $M'M$, which in turn equals the multiplicity of 2 as an eigenvalue of $MM' = K$. By Lemma 7.11, any eigenvector of K corresponding to 2 has no zero coordinate. Hence it has at most one eigenvector up to a scalar multiple. For, if x and y are linearly independent eigenvectors, both corresponding to the eigenvalue 2, then we may take a linear combination of x and y and get an eigenvector with a zero coordinate. Therefore the multiplicity of 2 as an eigenvalue of K is at most 1.   □

Lemma 7.13 *Let G be a bipartite graph. Then $\mathscr{L}_G$ is singular if and only if 2 is a Laplacian eigenvalue of G.*

Proof Since $M'M = 2I + A(\mathscr{L}_{(G)})$, $\mathscr{L}_{(G)}$ is singular if and only if 2 is a signless Laplacian eigenvalue of G. Since G is bipartite, by Lemma 7.7, K and L are similar, and therefore they have the same eigenvalues. This completes the proof.   □

Theorem 7.14 *Let T be a tree with n vertices where n is an odd integer. Then $\mathscr{L}_T$ is nonsingular.*

Proof If $\mathscr{L}_T$ is singular, then by Lemma 7.13, 2 is a Laplacian eigenvalue of T. Let $\phi(\lambda)$ be the characteristic polynomial of the Laplacian L of T. Since L is singular, $\phi(\lambda) = \lambda\phi_1(\lambda)$ for some polynomial ϕ_1 with integer coefficients. We may write $\phi_1(\lambda) = \lambda\phi_2(\lambda) + \phi_1(0)$, for some polynomial ϕ_2 with integer coefficients. Since 2 is a Laplacian eigenvalue of T, $\phi(2) = 0$. Hence $\phi_1(2) = 0$, and therefore $2\phi_2(2) = -\phi_1(0)$. Note that $\phi_1(0)$ is the coefficient of λ in $\phi(\lambda)$, which is the sum of the $(n-1) \times (n-1)$ principal minors of L. By the Matrix-Tree Theorem, each $(n-1) \times (n-1)$ principal minor of L is 1, and hence $2\phi_2(2) = -n$. Thus n must be even, which is a contradiction. This completes the proof.   □

Lemma 7.15 *Let G be a graph with an odd number of spanning trees and let k be an even integer. Then the multiplicity of k as an eigenvalue of $\mathscr{L}_G$ is at most 1.*

Proof Since $M'M = A(\mathscr{L}_G) + 2I$, the multiplicity of k as an eigenvalue of $\mathscr{L}_G$ equals the multiplicity of $k + 2$ as an eigenvalue of $M'M$. If $k \neq -2$, then the multiplicity of $k + 2$ as an eigenvalue of $M'M$ equals the multiplicity of $k + 2$ as an eigenvalue of $MM' = K$.

Let Q be the vertex-edge incidence matrix of the directed graph obtained by orienting each edge of G, and let $L = QQ'$ be the Laplacian. Since $Q \equiv M$ modulo 2, the determinants of $K(1|1)$ and $L(1|1)$ are equal modulo 2. Since G has an odd number of spanning trees, $\det L(1|1)$, and hence $\det K(1|1)$ are nonzero. If $X = K - (k+2)I$, then $\det X(1|1) \equiv \det K(1|1)$ modulo 2, and hence $\det X(1|1) \neq 0$. Thus the multiplicity of $k + 2$ as an eigenvalue of K is at most 1. (Here we have used the fact, which easily follows by interlacing, that if B is a symmetric, singular $n \times n$ matrix and if $\det B(1|1) \neq 0$, then the nullity of B is at most 1.)

Now suppose $k = -2$. Since G has an odd number of spanning trees, G is connected. By Lemma 7.5, the nullity of K is at most 1, and hence the multiplicity of $k + 2 = 0$ as an eigenvalue of K is at most 1. This completes the proof. $\qquad \square$

Corollary 7.16 *Let G be a graph with an odd number of spanning trees. Then the nullity of $\mathscr{L}_G$ is at most 1.*

Proof The result follows by setting $k = 0$ in Lemma 7.15. $\qquad \square$

Lemma 7.17 *Let G be a bipartite graph with an odd number vertices and an odd number of spanning trees. Then $\mathscr{L}_G$ is nonsingular.*

Proof If $\mathscr{L}_G$ is singular, then by Lemma 7.13, 2 is a Laplacian eigenvalue of G. Suppose $Ly = 2y$ for some nonzero vector y. We may take y to be an integer vector with its coordinates relatively prime. If y_i mod 2 equals 0, then $L(i|i)$, reduced modulo 2, must be singular. However, since G has an odd number of spanning trees, by the Matrix-Tree Theorem, $\det L(i|i)$ modulo 2 is nonzero, which is a contradiction. Thus y modulo 2 has no zero coordinate. Thus each y_i is an odd integer. Since $\mathbf{1}$ is an eigenvector of L, y must be orthogonal to $\mathbf{1}$, and hence $\sum_{i=1}^{n} y_i = 0$. This is a contradiction as each y_i is odd and n is odd. This completes the proof. $\quad \square$

Theorem 7.18 *Let G be a graph with $2^t s$ spanning trees where $t \geq 0, s > 0$ are integers and s is odd. Then the multiplicity of any even integer μ as an eigenvalue of K is at most $t + 1$.*

Proof Since $2^t s \neq 0$, G is connected. By Theorem 2.12, there exist unimodular matrices U and V such that

$$UAV = \mathsf{diag}(s_1, \ldots, s_{n-1}, 0),$$

where $s_1, \ldots, s_{n-1}$ are positive integers with $s_1 \cdots s_i = d_i$, where d_i is the greatest common divisor of all $i \times i$ minors of $L, i = 1, \ldots, n - 1$.

By the Matrix-Tree Theorem, $d_{n-1} = s_1 \cdots s_{n-1}$ equals the number of spanning trees of G. Since G has $2^t s$ spanning trees with s odd, at most t of the s_i are even. Therefore the rank of L, and hence the rank of K, over Z_2, the field of integers modulo 2, is at least $n - t - 1$. Recall that any symmetric matrix of rank r (over any field) has a principal $r \times r$ submatrix of full rank. Hence K has a nonsingular principal submatrix B of order $n - t - 1$. By interlacing, if an even integer μ is an eigenvalue of K with multiplicity at least $t + 2$, then any principal submatrix of K of order $n - t - 1$ has μ as an eigenvalue. So μ is an eigenvalue of B. It follows, as in the proof of Lemma 7.10, that μ divides $\det B$. Thus $\det B$ is even, which is a contradiction as B is nonsingular over Z_2. This completes the proof. $\qquad \square$

Corollary 7.19 *Let G be a graph with $2^t s$ spanning trees where $t \geq 0, s > 0$ are integers and s is odd. Then the nullity of $\mathscr{L}_G$ is at most $t + 1$.*

Note that Corollary 7.19 implies Corollary 7.16.

Theorem 7.20 *Let G be a graph with an odd number vertices and an odd number of spanning trees. If an even integer μ is an eigenvalue of the signless Laplacian K of G, then 4 divides μ.*

Proof Let G have n vertices. From Theorem 7.9, it follows that for $j = 1, \ldots, n-1$, $q_j = \ell_j + 4s_j$ for some integer s_j and $q_n = 4s_n$. This implies that $p_K(x) = p_L(x) + f(x)$, where $f(x)$ is an integer polynomial. Note that $\ell_{n-1} = (-1)^{n-1}$ times the number of spanning trees, which is an odd integer. It follows that if an even integer μ is congruent to 2 modulo 4, then so is $p_L(\mu)$. Thus $p_K(\mu)$ is nonzero for any even integer μ congruent to 2 modulo 4. Therefore if an even integer μ is an eigenvalue of K, then 4 divides μ. This completes the proof.   □

The next result is more general than Lemma 7.17.

Corollary 7.21 *Let G be a graph with an odd number of vertices and an odd number of spanning trees. Then $\mathscr{L}_G$ is nonsingular.*

Proof Let M be the $0 - 1$ incidence matrix of G and $K = MM'$. Since $M'M = A(\mathscr{L}_G) + 2I$, if $\mathscr{L}_G$ is singular, then 2 is an eigenvalue of $M'M$ and hence of K. This is a contradiction in view of Theorem 7.20 and the proof is complete.   □

Exercises

7.1. Let G be a connected block graph. Show that the following statements are equivalent: (i) G is the line graph of a tree (ii) G does not have the complete bipartite graph $K_{1,3}$ as an induced subgraph (iii) Any cut-vertex of G is contained in at most one block.

7.2. Let v be a pendant vertex in the graph G, adjacent to the vertex w. Show that the nullities of $G - v - w$ and G are the same.

7.3. Let G be a graph with a pendant vertex v, which is adjacent to w. Show that G is singular if and only if the nullity of $G - w$ is greater than 1. (We assume that G has at least 3 vertices.)

7.4. The star $S_{1,r}$ consists of a central vertex joined to r pendant vertices. The double star $T_{r,s}$ consists of two stars $S_{1,r}$ and $S_{1,s}$ joined together so that they share an edge. Let T be the double star $T_{r,s}, r \geq 3$. Determine $\det(A(\mathscr{L}_T))$.

7.5. Let T be the tree consisting of two stars $S_{1,r}, S_{1,s}; r, s \geq 3$, joined by an edge. Determine $\det(A(\mathscr{L}_T))$.

7.6. Find a tree whose line graph has no pendant vertices and the line graph is singular.

7.7. Let G be a graph with an odd number of vertices. Show that $\det(A(G))$ is even.

7.8. Let A be an $n \times n$ integral symmetric matrix with even entries on the diagonal. Show that if $n \equiv 0 \bmod 4$, then $\det A \equiv 0$ or $1 \bmod 4$, and if $n \equiv 2 \bmod 4$, then $\det A \equiv 0$ or $-1 \bmod 4$.

7.9. Let G be a graph with no perfect matching and let A be the adjacency matrix of G. Show that $\det A$ is even.

7.10. Let G be a graph with n vertices, let A be the adjacency matrix of G. Show that if $\det A = 1$, then $n \equiv 0 \bmod 4$, and if $\det A = -1$, then $n \equiv 2 \bmod 4$.

7.11. Show that the adjacency matrix of a tree is totally unimodular. Hence conclude that any rational eigenvalue of a nonsingular tree is ± 1.

7.12. Let T be a tree such that $\mathscr{L}_T$ is singular. Let $T + e$ be the tree obtained by adding a pendant edge to T. Then $\mathscr{L}_{T+e}$ is nonsingular. Let $T - v$ be the tree obtained from T be deleting the pendant vertex v. Then $\mathscr{L}_{T-v}$ is nonsingular.

7.13. Let T be a tree with n vertices. Show that $\det(A(\mathscr{L}_T)) = n$ modulo 2.

7.14. Let T be a tree with n vertices such that $\mathscr{L}_T$ is singular. Show that any $(n-2) \times (n-2)$ principal minor of $A(\mathscr{L}_T))$ is odd.

Section 7.1 is based on [Bap11]. The remaining sections are mainly based on [Bap11, Gho12, GS01]. Some further extensions are proved in [Gho13]; for example, it is shown that if a graph G has odd order and its number of spanning trees is not divisible by 4, then $A(\mathscr{L}_G)$ is nonsingular. An early reference is [GS01], where it was observed that the nullity of the line graph of a tree is at most 1. Exercises 7.8–7.10 are based on [AK07].

References and Further Reading

[AK07]   Akbari, S., Kirkland, S.J.: On unimodular graphs. Linear Algebra Appl. **421**, 3–15 (2007)

[Bap11] Bapat, R.B.: A note on singular line graphs. Bulletin Kerala Math. Assoc. **8**(2), 207–209 (2011)

[BS00]   Bapat, R.B., Souvik Roy: On the adjacency matrix of a block graph, Linear and Multilinear Algebra, to appear

[Gho12] Ebrahim Ghorbani: Spanning trees and line graph eigenvalues, arXiv:1201.3221v1 (2012)

[Gho13] Ebrahim Ghorbani: Spanning trees and line graph eigenvalues, arXiv:1201.3221v3 (2013)

[GS01]   Ivan Gutman, Irene Sciriha: On the nullity of line graphs of trees. Discrete Math. **232**, 35–45 (2001)

[Sci98] Irene Sciriha: On singular line graphs of trees. Congressus Numeratium **135**, 73–91 (1998)

Chapter 8
Algebraic Connectivity

Let G be a graph with $V(G) = \{1, \ldots, n\}$. Let L be the Laplacian of G and $0 = \lambda_1 \leq \lambda_2 \leq \cdots \leq \lambda_n$ be the eigenvalues of L. The second smallest eigenvalue, λ_2, is called the *algebraic connectivity* of G and is denoted by $\mu(G)$, or simply μ. Recall that if G is connected, then $\lambda_1 = 0$ is a simple eigenvalue, that is, it has algebraic multiplicity 1 and in that case $\mu > 0$. Conversely if $\mu = 0$, then G is disconnected. The complete graph K_n which may be regarded as "highly connected" has $\lambda_2 = \cdots = \lambda_n = n$. These observations justify the term "algebraic connectivity", introduced by Fiedler.

8.1 Preliminary Results

The following simple property of positive semidefinite matrices will be used.

Lemma 8.1 *Let B be an $n \times n$ positive semidefinite matrix. Then for any vector x of order $n \times 1$, $x'Bx = 0$ if and only if $Bx = 0$.*

Proof Note that $B = C'C$ for some $n \times n$ matrix C. Now $x'Bx = 0 \Rightarrow x'C'Cx = 0 \Rightarrow (Cx)'(Cx) = 0 \Rightarrow Cx = 0 \Rightarrow C'Cx = 0$, and hence $Bx = 0$. The converse is obvious. □

Let G be a connected graph with $V(G) = \{1, \ldots, n\}$. Let L be the Laplacian of G and μ be the algebraic connectivity. Let x be an eigenvector of L corresponding to μ. Then x is indexed by $V(G)$ and thus it gives a labeling of $V(G)$. That is, we label vertex i by x_i. We call vertex i positive, negative or zero according as $x_i > 0$, $x_i < 0$ or $x_i = 0$, respectively. Let

$$V^+ = \{i : x_i \geq 0\}, \quad V^- = \{i : x_i \leq 0\}.$$

With this notation we have the following basic result.

© Springer-Verlag London 2014
R.B. Bapat, *Graphs and Matrices*, Universitext,
DOI 10.1007/978-1-4471-6569-9_8

Theorem 8.2 *The subgraphs induced by V^+ and V^- are connected.*

Proof Since x is orthogonal to $\mathbf{1}$, the eigenvector of L corresponding to 0, then both V^+ and V^- are nonempty. We assume, without loss of generality, that $V^+ = \{1, \ldots, r\}$. Let, if possible, the subgraph of G induced by V^+ be disconnected and suppose, without loss of generality, that there is no edge from $\{1, \ldots, s\}$ to $\{s + 1, \ldots, r\}$. Then we may partition L as

$$L = \begin{bmatrix} L_{11} & 0 & L_{13} \\ 0 & L_{22} & L_{23} \\ L_{31} & L_{32} & L_{33} \end{bmatrix},$$

where L_{11} is $s \times s$ and L_{22} is $(r-s) \times (r-s)$. Partition x conformally and consider the equation

$$\begin{bmatrix} L_{11} & 0 & L_{13} \\ 0 & L_{22} & L_{23} \\ L_{31} & L_{32} & L_{33} \end{bmatrix} \begin{bmatrix} x^1 \\ x^2 \\ x^3 \end{bmatrix} = \mu \begin{bmatrix} x^1 \\ x^2 \\ x^3 \end{bmatrix}. \tag{8.1}$$

From (8.1) we have

$$L_{11}x^1 + L_{13}x^3 = \mu x^1. \tag{8.2}$$

Since $L_{13} \leq 0$ and $x^3 < 0$, we have $L_{13}x^3 \geq 0$. Since G is connected, L_{13} has a nonzero entry and hence $L_{13}x^3 \neq 0$. It follows from (8.2) that

$$(L_{11} - \mu I)x^1 \leq 0, \quad (L_{11} - \mu I)x^1 \neq 0. \tag{8.3}$$

From (8.3) we have

$$(x^1)'(L_{11} - \mu I)x^1 \leq 0. \tag{8.4}$$

We claim that $L_{11} - \mu I$ is not positive semidefinite. Indeed, if $L_{11} - \mu I$ is positive semidefinite, then $(x^1)'(L_{11} - \mu I)x^1 \geq 0$, which, together with (8.4) gives $(x^1)'(L_{11} - \mu I)x^1 = 0$. It follows by Lemma 8.1 that $(L_{11} - \mu I)x^1 = 0$. However, this contradicts (8.3) and hence we conclude that $L_{11} - \mu I$ is not positive semidefinite. Thus, L_{11} has an eigenvalue less than μ. A similar argument shows that L_{22} has an eigenvalue less than μ. Thus, the second smallest eigenvalue μ' of $\begin{bmatrix} L_{11} & 0 \\ 0 & L_{22} \end{bmatrix}$ is less than μ. However, by the interlacing theorem, $\mu \leq \mu'$, which is a contradiction. Therefore, the subgraph induced by V^+ is connected. It can similarly be proved that the subgraph induced by V^- is also connected. $\qquad\square$

An eigenvector corresponding to the algebraic connectivity is called a *Fiedler vector*.

Example 8.3 Consider the graph G:

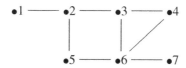

It may be verified that the algebraic connectivity of G is 0.5926. A Fiedler vector, rounded to two decimal places, is given by

$$[0.71, 0.29, -0.06, -0.21, 0.04, -0.23, -0.56]'.$$

Thus, vertices 1, 2 and 5 are positive and they induce a connected subgraph. Vertices 3, 4, 6 and 7 are negative and they induce a connected subgraph as well.

8.2 Classification of Trees

We now consider the case of trees in greater detail. Let T be a tree with $V(T) = \{1, \ldots, n\}$ and the edge set $E(T) = \{e_1, \ldots, e_{n-1}\}$. Let L be the Laplacian of T and μ be the algebraic connectivity. Let x be an eigenvector of L corresponding to μ. We refer to x as a Fiedler vector of L. First, suppose that x has no zero coordinate. Then

$$V^+ = \{i : x_i > 0\}, \quad V^- = \{i : x_i < 0\}$$

give a partition of $V(T)$. By Theorem 8.2, the subgraphs induced by V^+ and V^- must be connected and then, clearly, they must both be trees. Recall that a vertex i is positive or negative according as $x_i > 0$ or $x_i < 0$, respectively. Then there must be precisely one edge such that one of its end-vertices is positive and the other negative. Such an edge is called a *characteristic edge* (with respect to x). Any other edge has either both its end-vertices positive or both negative.

We turn to the case where a Fiedler vector has a zero coordinate. This case requires a closer analysis by means of some subtle properties of interlacing of eigenvalues. Note that $Lx = \mu x$ implies that

$$\sum_{j \sim i} x_j = (d_i - \mu)x_i, \tag{8.5}$$

where d_i is the degree of i. If $x_i = 0$ then (8.5) implies that either $x_j = 0$ for all j adjacent to i or i is adjacent to a positive vertex as well as a negative vertex. A zero vertex is called a *characteristic vertex* (with respect to x) if it is adjacent to a positive vertex and a negative vertex. It is evident from (8.5) that a pendant vertex cannot be a characteristic vertex. Our goal is to prove the interesting fact that corresponding to any Fiedler vector a tree has at most one characteristic vertex.

We first develop some preliminary results. If A is an $n \times n$ symmetric matrix, then $p_+(A)$, $p_-(A)$ and $p_0(A)$ will denote, respectively, the number of positive, negative and zero eigenvalues of A. Thus, $p_+(A) + p_-(A) + p_0(A) = n$. The 3-tuple $(p_+(A), p_-(A), p_0(A))$ is called the *inertia* of A.

Lemma 8.4 *Let B be a symmetric $n \times n$ matrix and let c be a vector of order $n \times 1$. Suppose there exists a vector u such that $Bu = 0$ and $c'u \neq 0$. Let*

$$A = \begin{bmatrix} B & c \\ c' & d \end{bmatrix},$$

where d is a real number. Then

$$p_+(A) = p_+(B) + 1, \quad p_-(A) = p_-(B) + 1 \quad and \quad p_0(A) = p_0(B) - 1.$$

Proof First note that $u \neq 0$ since $c'u \neq 0$. Then $Bu = 0$ implies that B is singular and 0 is an eigenvalue of B. If c were in the column space of B, then c would be equal to By for some vector y. Then $u'c = u'By = 0$, since $Bu = 0$. This is a contradiction since $c'u \neq 0$. Therefore, c is not in the column space of B. Thus,

$$\text{rank}(A) = \text{rank}[B, c] + 1 = \text{rank}(B) + 2. \tag{8.6}$$

Since the rank of an $m \times m$ symmetric matrix is m minus the multiplicity of the zero eigenvalue, it follows from (8.6) that $p_0(A) = p_0(B) - 1$. By the interlacing theorem, $p_+(B) \leq p_+(A) \leq p_+(B) + 1$ and $p_-(B) \leq p_-(A) \leq p_-(B) + 1$. These conditions together imply that $p_+(A) = p_+(B) + 1$ and $p_-(A) = p_-(B) + 1$. That completes the proof. □

Corollary 8.5 *Let A be a symmetric matrix partitioned as*

$$A = \begin{bmatrix} A_{11} & A_{12} \\ A_{21} & A_{22} \end{bmatrix},$$

where A_{11} and A_{22} are square. Let u be a vector such that $A_{11}u = 0$ and $A_{21}u \neq 0$. Then $p_-(A) \geq p_-(A_{11}) + 1$.

Proof Since $A_{21}u \neq 0$, there exists a column c of A_{12} such that $c'u \neq 0$. Let d be the diagonal entry of A_{22} corresponding to the column c of A_{12}. Consider the matrix

$$X = \begin{bmatrix} A_{11} & c \\ c' & d \end{bmatrix}.$$

By Lemma 8.4, $p_-(X) = p_-(A_{11}) + 1$. Also, since X is a principal submatrix of A, by the interlacing theorem, $p_-(A) \geq p_-(X)$. It follows that $p_-(A) \geq p_-(A_{11}) + 1$. □

In Theorem 6.3, Chap. 6, we proved the main aspects of Perron–Frobenius theory confining ourselves to adjacency matrices. The theorem can be proved, essentially by the same method, for any nonnegative, "irreducible" matrix. Here we do not yet need the theorem in its full generality, however we do need it for a small modification of the adjacency matrix. The result is stated next. The proof is along the lines of that of Theorem 6.3.

Theorem 8.6 *Let G be a connected graph with n $\geq$ 2 vertices, and let A be the adjacency matrix of G. Let E $\geq$ 0 be a diagonal matrix. Then the following assertions hold:*

(i) *E + A has an eigenvalue $\lambda > 0$ and an associated eigenvector $x > 0$.*
(ii) *For any eigenvalue $\mu \neq \lambda$ of E + A, $-\lambda \leq \mu < \lambda$.*
(iii) *If u is an eigenvector of E + A for the eigenvalue λ, then $u = \alpha x$ for some α.*

We will refer to the eigenvalue λ in (i) of Theorem 8.6 as the Perron eigenvalue of $E + A$.

Corollary 8.7 *Let G be a connected graph with n vertices and let A be the adjacency matrix of G. Let E be a diagonal matrix of order n and let $\tau_1 \leq \tau_2 \cdots \leq \tau_n$ be the eigenvalues of E − A. Then the algebraic multiplicity of τ_1 is 1 and there is a positive eigenvector of E − A corresponding to τ_1.*

Proof Let $B = kI - (E - A)$, where $k > 0$ is sufficiently large so that $kI - E \geq 0$. The eigenvalues of B are $k - \tau_1 \geq k - \tau_2 \cdots \geq k - \tau_n$. Since $B = (kI - E) + A$, by Theorem 8.6 $k - \tau_1$, which is the Perron eigenvalue of B, has algebraic multiplicity 1 and there is a positive eigenvector corresponding to the same. It follows that τ_1, as an eigenvalue of $E - A$, has algebraic multiplicity 1 with an associated positive eigenvector. $\square$

For a symmetric matrix B, let $\tau(B)$ denote the least eigenvalue of B.

Theorem 8.8 *Let T be a tree with $V(T) = \{1, \ldots, n\}$. Let L be the Laplacian of T and μ the algebraic connectivity. Let x be a Fiedler vector and suppose n is a characteristic vertex. Let $T_1, \ldots, T_k$ be the components of $T \setminus \{n\}$. Then for any $j = 1, \ldots, k$, the vertices of $V(T_j)$ are either all positive, all negative or all zero.*

Proof Recall that since n is a characteristic vertex, $x_n = 0$ and n is adjacent to a positive as well as a negative vertex. As observed earlier, n cannot be a pendant vertex and hence $k \geq 2$. Partition L and x conformally so that $Lx = \mu x$ is expressed as

$$\begin{bmatrix} L_1 & 0 & \cdots & 0 \\ 0 & L_2 & \cdots & 0 \\ \vdots & \vdots & \ddots & \vdots & w \\ 0 & 0 & \cdots & L_k \\ & & w' & & d_n \end{bmatrix} \begin{bmatrix} x^1 \\ x^2 \\ \vdots \\ x^k \\ 0 \end{bmatrix} = \mu \begin{bmatrix} x^1 \\ x^2 \\ \vdots \\ x^k \\ 0 \end{bmatrix}, \qquad (8.7)$$

where L_j is the submatrix of L corresponding to vertices in T_j, $j = 1, \ldots, k$, and d_n is the degree of n. We must show that for $j = 1, \ldots, k$, $x^j > 0$, $x^j < 0$ or $x^j = 0$.

Suppose $x^1 \neq 0$, $x^2 \neq 0$. From (8.7), $L_j x^j = \mu x^j$, $j = 1, 2$. Thus, μ is an eigenvalue of L_j, $j = 1, 2$, and hence $\tau(L_j) \leq \mu$, $j = 1, 2$. First suppose $\tau(L_1) \neq \tau(L_2)$ and we first consider the case $\tau(L_1) < \tau(L_2)$. Let $L(n, n)$ be the principal submatrix of L obtained by deleting the row and the column n. By Corollary 8.7 there exists a vector $u > 0$ such that $L_2 u = \tau(L_2)u$. Augment u suitably by zeros to get the vector $\tilde{u} = [0, u, 0, \ldots, 0]'$, which satisfies $L(n, n)\tilde{u} = \tau(L_2)\tilde{u}$. There is a vertex of T_2 adjacent to n and hence $\tilde{u}'w \neq 0$. By Corollary 8.5 it follows that

$$p_-(L - \tau(L_2)I) \geq p_-(L(n, n) - \tau(L_2)I) + 1. \tag{8.8}$$

Since $\tau(L_1) < \tau(L_2)$, then $p_-(L(n, n) - \tau(L_2)I) \geq 1$, and it follows from (8.8) that $p_-(L - \tau(L_2)I) \geq 2$. We conclude that $\mu < \tau(L_2)$, which contradicts the earlier observation that $\tau(L_2) \leq \mu$. Hence, it is not possible that $\tau(L_1) < \tau(L_2)$. By a similar argument we can show that $\tau(L_2)$ cannot be less than $\tau(L_1)$.

Now suppose $\tau(L_1) = \tau(L_2) \leq \mu$. Then $\begin{bmatrix} L_1 & 0 \\ 0 & L_2 \end{bmatrix}$ has at least two eigenvalues not exceeding μ. By the interlacing theorem, L must have two eigenvalues not exceeding $\tau(L_1)$. It follows that $\tau(L_1) = \tau(L_2) = \mu$. By Corollary 8.7 it follows that $x^j > 0$ or $x^j < 0$ for $j = 1, 2$. A similar argument shows that for $j = 3, \ldots, k$, if $x_j \neq 0$ then either $x^j > 0$ or $x^j < 0$. That completes the proof. $\square$

Corollary 8.9 *Let T be a tree with $V(T) = \{1, \ldots, n\}$. Let L be the Laplacian of T and μ the algebraic connectivity. Let x be a Fiedler vector. Then T has at most one characteristic vertex with respect to x.*

Proof Suppose $i \neq j$ are both characteristic vertices with respect to x. Then $x_i = x_j = 0$. By Theorem 8.8 all vertices of the component of $T \setminus \{i\}$ that contains j are zero vertices. Then j cannot be adjacent to a nonzero vertex and thus it cannot be a characteristic vertex. $\square$

Let A be a symmetric $n \times n$ matrix. We may associate a graph G_A with A as follows. Set $V(G_A) = \{1, \ldots, n\}$. For $i \neq j$, vertices i and j are adjacent if and only if $a_{ij} \neq 0$.

Lemma 8.10 *Let A be a symmetric $n \times n$ matrix such that G_A is a tree, and suppose $A\mathbf{1} = 0$. Then $\mathrm{rank}(A) = n - 1$.*

Proof We prove the result by induction on n. The proof is easy when $n = 2$. Assume the result to be true for matrices of order $n - 1$. We assume, without loss of generality, that vertex n is pendant and is adjacent to $n - 1$. Let z be a vector such that $Az = 0$. Then the nth equation gives $a_{n-1,n}z_{n-1} + a_{nn}z_n = 0$. Since $a_{n-1,n} = -a_{nn} \neq 0$, it follows that $z_{n-1} = z_n$. As usual, let $A(n, n)$ be the submatrix obtained by deleting row and column n of A. Also, let $z(n)$ be the vector obtained by deleting the last coordinate of z. The first $n - 1$ equations from $Az = 0$ give

$$\left(A(n, n) + \begin{bmatrix} 0 & \cdots & 0 \\ \vdots & \ddots & \vdots \\ 0 & \cdots & -a_{nn} \end{bmatrix} \right) z(n) = 0. \tag{8.9}$$

Let $B = A(n, n) - \mathsf{diag}(0, \ldots, 0, a_{nn})$, which is the matrix on the left side of (8.9). Note that G_B is the tree $T \setminus \{n\}$ and $B\mathbf{1} = 0$. By, induction assumption it follows that $\mathsf{rank}(B) = n - 2$ and therefore $z(n)$ must be a scalar multiple of $\mathbf{1}$. It follows that z is a scalar multiple of $\mathbf{1}$ and hence $\mathsf{rank}(A) = n - 1$. □

Theorem 8.11 *Let G be a tree with $V(G) = \{1, \ldots, n\}$. Let L be the Laplacian of G and μ the algebraic connectivity. Suppose there exists a Fiedler vector with no zero coordinate. Then μ has algebraic multiplicity 1.*

Proof Let $Ly = \mu y$ where $y_i \neq 0, i = 1, \ldots, n$. Let $E = \mathsf{diag}(y_1, \ldots, y_n)$ and let $C = E(L - \mu I)E$. Then G_C is a tree and $C\mathbf{1} = 0$. It follows by Lemma 8.10 that $\mathsf{rank}(C) = n - 1$. Then $\mathsf{rank}(L - \mu I)$ is $n - 1$ as well, and hence μ has algebraic multiplicity 1. □

Let T be a tree with $V(T) = \{1, \ldots, n\}$. Let L be the Laplacian of T and μ the algebraic connectivity. Let x be a Fiedler vector and suppose x has no zero coordinate. Then by Theorem 8.11, μ has algebraic multiplicity 1, and hence any other Fiedler vector must be a scalar multiple of x. Thus, in this case there is an edge of T that is the characteristic edge with respect to *every* Fiedler vector. An analogous result holds for a characteristic vertex as well, as seen in the next result.

Theorem 8.12 *Let T be a tree with $V(T) = \{1, \ldots, n\}$. Let L be the Laplacian of T and μ the algebraic connectivity. Let x and y be Fiedler vectors. Then a vertex i is a characteristic vertex with respect to x if and only if it is a characteristic vertex with respect to y.*

Proof At the outset we note a consequence of Theorem 8.8, which will be used. If x is a Fiedler vector of a tree and has a zero coordinate, then for any vertices i, j and k of the tree such that j is on the $i - k$ path, if $x_i = x_k = 0$ then $x_j = 0$.

We turn to the proof. If μ has algebraic multiplicity 1 then x is a scalar multiple of y and the result is obvious. So, suppose μ has algebraic multiplicity greater than 1, and let

$$V_0 = \{j \in V(T) : z_j = 0, \text{ for any Fiedler vector } z\}.$$

If $V_0 = \phi$ then for each vertex j we can find a Fiedler vector z^j such that the jth coordinate of z^j is nonzero. Then there must be a vector z with no zero entry that is a linear combination of z^j, $j = 1, \ldots, n$. Note that z is a Fiedler vector, contradicting Theorem 8.11. Therefore, $V_0 \neq \phi$.

There must be a vertex $k \in V_0$ that is adjacent to a vertex not in V_0. Suppose there are two vertices $k_1, k_2 \in V_0$ adjacent to vertices not in V_0. Specifically, suppose k_1 is adjacent to $\ell_1 \notin V_0$ and k_2 is adjacent to $\ell_2 \notin V_0$. Then there are Fiedler vectors w^1 and w^2 such that the k_i-coordinate of w^i is zero, while the ℓ_i-coordinate of w^i is nonzero, $i = 1, 2$. We may take a linear combination w of w^1 and w^2, which then is a Fiedler vector, with respect to which both k_1 and k_2 are characteristic vertices, contradicting Corollary 8.9. Hence, $k \in V_0$ is the unique vertex adjacent to a vertex not in V_0.

We claim that k is the characteristic vertex with respect to any Fiedler vector. Suppose $i \neq k$ is the characteristic vertex with respect to the Fiedler vector x. There must be a vertex j adjacent to i such that $x_j \neq 0$. Since $i \notin V_0$, there is a Fiedler vector y such that $y_i \neq 0$. Since $x_i = x_k = 0$ and $x_j \neq 0$, by the structure implied by Theorem 8.8, i is on the $j - k$ path. It follows by the observation in the beginning that $y_j \neq 0$.

We may take a linear combination z of x and y satisfying $z_j = 0$ and $z_i \neq 0$. However, $z_k = 0$, which again contradicts the observation in the beginning since i is on the $j - k$ path. We conclude that k is the characteristic vertex with respect to any Fiedler vector.   □

We are now in a position to describe a classification of trees. Let T be a tree with $V(T) = \{1, \ldots, n\}$. We say that T is of Type I if it has a characteristic vertex with respect to any Fiedler vector, while T is said to be of Type II if it has a characteristic edge with respect to any Fiedler vector. As discussed earlier, neither the characteristic vertex nor the characteristic edge depend on the particular Fiedler vector. Note that every tree must be of one of the two types. A tree cannot be both Type I and Type II. Indeed, in that case μ must have algebraic multiplicity at least 2 and then, by Theorem 8.11, it cannot have a Fiedler vector with all coordinates nonzero, a contradiction.

It must be remarked that if μ has algebraic multiplicity greater than 1 then T is necessarily of Type I. However, the converse is not true. If T is the path on 3 vertices then the central vertex is a characteristic vertex, although the algebraic multiplicity of $\mu = 1$ is 1.

8.3 Monotonicity Properties of Fiedler Vector

The coordinates of a Fiedler vector exhibit a monotonicity property in the case of both Type I and Type II trees. We first prove a preliminary result, which will be used in proving the monotonicity.

Lemma 8.13 *Let T be a tree with $V(T) = \{1, \ldots, n\}$. Let L be the Laplacian of T. Let λ be an eigenvalue of L and let z be a corresponding eigenvector. Let $e = \{i, j\}$ be an edge of T. Then*

$$z_i - z_j = -\lambda \sum_k z_k$$

where the summation is over all the vertices k in the component of $T \setminus \{e\}$ that contains j.

Proof We assume, after a relabeling, that the edge e has endpoints s and $s + 1$, and furthermore, the two components of $T \setminus \{e\}$ have vertex sets $\{1, \ldots, s\}$ and $\{s + 1, \ldots, n\}$. Let u be the vector of order $n \times 1$ with $u_i = 1$, $i = 1, \ldots, s$, and $u_i = 0$, $i = s + 1, \ldots, n$. Note that

$$u'L = [0, \ldots, 0, 1, -1, 0, \ldots, 0],$$

where the 1 and the -1 occur at positions s and $s + 1$, respectively. Therefore, $u'Lz = z_s - z_{s+1}$. Hence, from $u'Lz = \lambda u'z$ we conclude that

$$z_s - z_{s+1} = \lambda \sum_{k=1}^{s} z_k. \tag{8.10}$$

Since z is orthogonal to $\mathbf{1}$, the expression on the right side of (8.10) equals $-\lambda \sum_{k=s+1}^{n} z_k$. This completes the proof. $\qquad\square$

The following result has been partly proved in the earlier discussion.

Theorem 8.14 *Let T be a tree with $V(T) = \{1, \ldots, n\}$. Let L be the Laplacian of T and μ the algebraic connectivity. Let x be a Fiedler vector. Then one of the following cases occur.*

(i) *No entry of x is zero. In this case there is a unique edge $e = \{i, j\}$ such that $x_i > 0$ and $x_j < 0$. Further, along any path in T that starts at i and does not contain j, the entries of x increase, while along any path in T that starts at j and does not contain i, the entries of x decrease.*

(ii) *Some entry of x is zero. In this case the subgraph of T induced by the zero vertices is connected. There is a unique vertex k such that $x_k = 0$ and k is adjacent to a nonzero vertex. Further, along any path in T that starts at k, the entries of x either increase or decrease.*

Proof (i) First suppose no entry of x is zero. In this case, by Theorem 8.2 there is a unique edge (the characteristic edge) $e = \{i, j\}$ such that $x_i > 0$ and $x_j < 0$. Consider any edge $f = \{u, v\}$ on a path that starts at i and does not contain j. Assume that u is closer to i than v. By Lemma 8.13,

$$x_u - x_v = -\mu \sum_{k} x_k, \tag{8.11}$$

where the summation is over all vertices k in the component of $T \setminus \{f\}$ that contains v. Note that all the vertices in this component are positive and hence it follows from (8.11) that $x_u < x_v$. Thus, along any path in T that starts at i and does not contain j, the entries of x increase. The second part of (i) is proved similarly.

(ii) Suppose x has a zero coordinate. By Theorem 8.8 there is a unique vertex (the characteristic vertex) k such that $x_k = 0$ and k is adjacent to a nonzero vertex.

Further, the vertices in any component of $T \setminus \{k\}$ are either all positive, all negative or all zero. It follows that the subgraph of T induced by the zero vertices is connected. The proof of the second part is similar to the one given for (i). $\qquad\square$

8.4 Bounds for Algebraic Connectivity

The following representation for the second smallest eigenvalue of a symmetric matrix will be used. It is easily derived from the spectral theorem.

Lemma 8.15 *Let A be a symmetric $n \times n$ matrix with eigenvalues $\lambda_1 \geq \cdots \geq \lambda_{n-1} \geq \lambda_n$. Let u be an eigenvector of A corresponding to λ_n. Then*

$$\lambda_{n-1} = \min \left\{ \frac{x'Ax}{x'x} \right\},$$

where the minimum is taken over all nonzero vectors x, orthogonal to u.

We introduce some notation. Let G be a connected graph with $V(G) = \{1, \ldots, n\}$. If $i, j \in V(G)$, then as usual the distance between i and j, denoted $d(i, j)$, is defined to be the length (that is, the number of edges) in the shortest (ij)-path. We set $d(i, i) = 0, i = 1, \ldots, n$. If $V_1, V_2 \subset V(G)$ are nonempty sets then define

$$d(V_1, V_2) = \min\{d(i, j) : i \in V_1, j \in V_2\}.$$

If $V_1 = \{i\}$ we write $d(V_1, V_2)$ as $d(i, V_2)$.

Theorem 8.16 *Let G be a connected graph with $V(G) = \{1, \ldots, n\}$. Let V_1 and V_2 be nonempty disjoint subsets of $V(G)$, and let G_1 and G_2 be the subgraphs induced by V_1 and V_2, respectively. Let L be the Laplacian of G and μ the algebraic connectivity. Then*

$$\mu \leq \frac{1}{d(V_1, V_2)^2} \left(\frac{1}{|V_1|} + \frac{1}{|V_2|} \right) (|E(G)| - |E(G_1)| - |E(G_2)|).$$

Proof Let

$$g(i) = \frac{1}{|V_1|} - \frac{1}{d(V_1, V_2)} \left(\frac{1}{|V_1|} + \frac{1}{|V_2|} \right) \min\{d(i, V_1), d(V_1, V_2)\},$$

$i = 1, \ldots, n$. Note that if $i \in V_1$ then $g(i) = \frac{1}{|V_1|}$, and if $i \in V_2$ then $g(i) = -\frac{1}{|V_2|}$. Also, if $i \sim j$ then $|d(i, V_1) - d(j, V_2)| \leq 1$ and hence

$$|g(i) - g(j)| \leq \frac{1}{d(V_1, V_2)} \left(\frac{1}{|V_1|} + \frac{1}{|V_2|} \right). \qquad (8.12)$$

Let $\bar{g} = \frac{1}{n}\sum_{j \in V(G)} g(j)$, and let $f(i) = g(i) - \bar{g}$, $i = 1, \ldots, n$. Let f be the vector of order $n \times 1$ with the ith component $f(i)$, $i = 1, \ldots, n$. Then $f'\mathbf{1} = 0$. It follows from Lemma 4.3 (iii) that

$$f'Lf = \sum_{i \sim j}(f(i) - f(j))^2 = \sum_{i \sim j}(g(i) - g(j))^2. \tag{8.13}$$

If i and j are both in V_1 or are both in V_2, then $g(i) = g(j)$. If $\{i, j\}$ is any edge not in $E(G_1) \cup E(G_2)$, then by (8.12),

$$(g(i) - g(j))^2 \le \frac{1}{d(V_1, V_2)^2}\left(\frac{1}{|V_1|} + \frac{1}{|V_2|}\right)^2. \tag{8.14}$$

We conclude from (8.13) and (8.14) that

$$f'Lf \le \frac{1}{d(V_1, V_2)^2}\left(\frac{1}{|V_1|} + \frac{1}{|V_2|}\right)^2 (|E(G)| - |E(G_1)| - |E(G_2)|). \tag{8.15}$$

Observe that

$$\begin{aligned}
f'f &= \sum_{i \in V(G)} f(i)^2 \\
&\ge \sum_{i \in V_1 \cup V_2} f(i)^2 \\
&= |V_1|\left(\frac{1}{|V_1|} - \bar{g}\right)^2 + |V_2|\left(\frac{1}{|V_2|} + \bar{g}\right)^2 \\
&\ge \frac{1}{|V_1|} + \frac{1}{|V_2|}.
\end{aligned} \tag{8.16}$$

Since $f'\mathbf{1} = 0$, it follows from Lemma 8.15 that

$$\mu f'f \le f'Lf. \tag{8.17}$$

The result follows from (8.15), (8.16) and (8.17). □

We indicate some consequences of Theorem 8.16.

Corollary 8.17 *Let G be a connected graph with $V(G) = \{1, \ldots, n\}$. Let L be the Laplacian of G and μ the algebraic connectivity. Let δ be the minimum vertex degree in G. Then $\mu \le \frac{n}{n-1}\delta$.*

Proof Let i be a vertex of degree δ. Let $V_1 = \{i\}$ and $V_2 = V(G) \setminus \{i\}$. Then $d(V_1, V_2) = 1$. Using the notation of Theorem 8.16 we see that $|E(G)| - |E(G_1)| - |E(G_2)| = \delta$. The result easily follows by an application of Theorem 8.16. □

Corollary 8.18 *Let G be a connected, k-regular graph with n vertices and with algebraic connectivity μ. Let H be an induced subgraph of G with p vertices. Then the average degree of a vertex in H is at most $\frac{p\mu}{n} + k - \mu$.*

Proof Let $V_1 = V(H)$, $V_2 = V(G) \setminus V(H)$. Then $d(V_1, V_2) = 1$. Applying Theorem 8.16 we see that the total number of edges between the vertices of H and the vertices not in H is at least $\mu \frac{p(n-p)}{n}$. Thus, the sum of the degrees (in H) of the vertices in H is at most

$$kp - \mu \frac{p(n-p)}{n} = p \left(\frac{p\mu}{n} + k - \mu \right).$$

Hence, the average degree of a vertex in H is at most $\frac{p\mu}{n} + k - \mu$.   □

Let G be a connected graph with $V(G) = \{1, \ldots, n\}$. Let V_1 be a nonempty subset of $V(G)$ and let $b(V_1)$ be the number of edges with precisely one endpoint in V_1. The minimum value of $\frac{b(V_1)}{|V_1|}$ taken over all V_1 with $|V_1| \le \frac{n}{2}$ is called the *isoperimetric number* of G. It is an easy consequence of Theorem 8.16 that the isoperimetric number is at least $\frac{\mu}{2}$.

We conclude with yet another inequality that can be derived from Theorem 8.16.

Corollary 8.19 *Let G be a connected graph with $V(G) = \{1, \ldots, n\}$. Let V_1 and V_2 be nonempty disjoint subsets of $V(G)$ and let G_1 and G_2 be the subgraphs induced by V_1 and V_2 respectively. Let L be the Laplacian of G and μ the algebraic connectivity. Let Δ be the maximum vertex degree in G. Suppose $d(V_1, V_2) > 1$. Then*

$$\mu \le \frac{\Delta}{d(V_1, V_2)^2} \cdot \frac{n}{|V_1||V_2|} (n - |V_1| - |V_2|).$$

Proof Since $d(V_1, V_2) > 1$, every edge in $E(G) \setminus (E(G_1) \cup E(G_2))$ is incident with at least one of the $n - |V(G_1)| - |V(G_2)|$ vertices of the set $V(G) \setminus (V(G_1) \cup V(G_2))$. Thus,

$$|E(G)| - |E(G_1)| - |E(G_2)| \le \Delta(n - |V(G_1)| - |V(G_2)|). \tag{8.18}$$

By Theorem 8.16 and (8.18) we get

$$\mu \le \frac{1}{d(V_1, V_2)^2} \left(\frac{1}{|V_1|} + \frac{1}{|V_2|} \right) (|E(G)| - |E(G_1)| - |E(G_2)|) \tag{8.19}$$

$$\le \frac{1}{d(V_1, V_2)^2} \left(\frac{1}{|V_1|} + \frac{1}{|V_2|} \right) (n - |V_1| - |V_2|) \Delta \tag{8.20}$$

$$\le \frac{\Delta}{d(V_1, V_2)^2} \cdot \frac{n}{|V_1||V_2|} (n - |V_1| - |V_2|) \tag{8.21}$$

and the proof is complete.   □

In the next result we give an inequality between the algebraic connectivity of a graph and that of an induced subgraph.

Theorem 8.20 *Let G be a connected graph with $V(G) = \{1, \ldots, n\}$. Let L be the Laplacian of G and μ the algebraic connectivity. Let V_1 and V_2 be nonempty disjoint subsets of $V(G)$ with $V_1 \cup V_2 = V(G)$. Let H be the subgraph induced by V_1 and let μ_1 be the algebraic connectivity of H. Then*

$$\mu \leq \mu_1 + |V_2|.$$

Proof Let x be a unit Fiedler vector of H. Augment x by zeros resulting in a vector of order $n \times 1$, which we denote by z. Then z is also a unit vector and $z'1 = 0$. It follows by Lemma 8.15 and Lemma 4.3 that

$$\mu \leq z'Lz = \sum_{i \sim j} (z_i - z_j)^2.$$

Decompose the preceding sum into three parts: edges (i, j) with no endpoint in V_1, one endpoint in V_1 and both endpoints in V_1. Ignore the first sum and observe that the second sum is bounded above by $|V_2|$. Finally, the third sum equals μ_1 and the result follows. □

Exercises

1. Determine the algebraic connectivity of the star $K_{1,n-1}$.
2. Let G be a connected graph and let x be a Fiedler vector. If $x_i > 0$ then show that there exists a vertex $j \sim i$ such that $x_i > x_j$.
3. Let x be a Fiedler vector of a unicyclic graph with vertex set $\{1, \ldots, n\}$ and suppose x has no zero coordinate. Show that there are at most two edges such that their end-vertices are of different signs.
4. Let P_n be the path with n vertices, where $n \geq 3$ is odd. Show that the central vertex is a characteristic vertex.
5. Let G be a connected graph with $n = 2m$ vertices. Let V_1 and V_2 be disjoint subsets of $V(G)$ with $|V_1| = |V_2| = m$. Let μ be the algebraic connectivity of G. Show that the number of edges of G with one endpoint in V_1 and the other in V_2 is at least $\frac{\mu m}{2}$. Show that equality is attained for the n-cube Q_n, $n \geq 2$, by taking suitable V_1 and V_2.
6. Let G be a connected graph with $n = 3m$ vertices. Let V_1, V_2 and V_3 be disjoint subsets of $V(G)$ with $|V_1| = |V_2| = |V_3| = m$. Let μ be the algebraic connectivity of G. Show that the number of edges of G with one endpoint in V_i and the other in V_j for some $i \neq j$ is at least $m\mu$.

7. Let G be a connected graph with $V(G) = \{1, \ldots, n\}$. Let μ be the algebraic connectivity of G. Let $V_1 \subset V(G)$ and suppose the graph induced by $V(G) \setminus V_1$ is disconnected. Show that $\mu \leq |V_1|$.

8. Show that the algebraic connectivity does not exceed the minimum vertex degree. (This statement is stronger than Corollary 8.17

9. Show that the algebraic connectivity of P_n, the path on n vertices, does not exceed $\frac{12}{n^2-1}$.

10. Is it true that the algebraic connectivity necessarily decreases when a vertex is deleted?

The basic theory outlined in Sects. 7.1–7.3 is due to Fiedler [F73, F75]. We have also incorporated results and proof techniques from [BP98, BBT92, GR01, M87] in these sections. Section 7.4 is based on [AM85]. Bounds for the isoperimetric number are important in the study of expander graphs; see [CR09].

References and Further Reading

[AM85]   Alon, N., Milman, V.D.: λ_1, isoperimetric inequalities for graphs and superconcentrators. J. Comb. Theor. Ser. B **38**, 73–88 (1985)

[BP98]   Bapat, R.B., Pati, S.: Algebraic connectivity and the characteristic set of a graph. Linear Multilinear Algebra **45**, 247–273 (1998)

[BBT92] Biggs, N.L., Brightwell, G.R., Tsoubelis, D.: Theoretical and practical studies of a competitive learning process. Netw. Comput. Neural Syst. **3**(3), 285–301 (1992)

[CR09]   Cioabă, S.M., Ram Murty, M.: A First Course in Graph Theory and Combinatorics, Texts and Readings in Mathematics 55. Hindustan Book Agency, New Delhi (2009)

[F73]   Fiedler, M.: Algebraic connectivity of graphs. Czech. Math. J **23**(98), 298–305 (1973)

[F75]   Fiedler, M.: Eigenvalues of acyclic matrices. Czech. Math. J **25**(100), 607–618 (1975)

[GR01]   Godsil, C., Royle, G.: Algebraic Graph Theory. Springer, New York (2001)

[M87]   Merris, R.: Characteristic vertices of trees. Linear Multilinear Algebra **22**, 115–131 (1987)

Chapter 9
Distance Matrix of a Tree

Let G be a connected graph with $V(G) = \{1, \ldots, n\}$. Recall that the distance $d(i, j)$ between the vertices i and j of G is the length of a shortest path from i to j. The distance matrix $D(G)$ of G is an $n \times n$ matrix with its rows and columns indexed by $V(G)$. For $i \neq j$, the (i, j)-entry d_{ij} of G is set equal to $d(i, j)$. Also, $d_{ii} = 0$, $i = 1, \ldots, n$. We will often denote $D(G)$ simply by D. Clearly, D is a symmetric matrix with zeros on the diagonal. The distance, as a function on $V(G) \times V(G)$, satisfies the triangle inequality. Thus, for any vertices i, j and k,

$$d(i, k) \leq d(i, j) + d(j, k).$$

The proof is easy. If $d(i, j)$ is the length of the (ij)-path $\mathscr{P}_1$ and $d(j, k)$ is the length of the (jk)-path $\mathscr{P}_2$, then $\mathscr{P}_1 \cup \mathscr{P}_2$ contains an (ik)-path. Therefore, the length of a shortest (ik)-path cannot exceed the sum $d(i, j) + d(j, k)$.

Example 9.1 Consider the tree

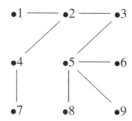

The distance matrix of the tree is given by

© Springer-Verlag London 2014
R.B. Bapat, *Graphs and Matrices*, Universitext,
DOI 10.1007/978-1-4471-6569-9_9

$$\begin{bmatrix}
0 & 1 & 2 & 2 & 3 & 4 & 3 & 4 & 4 \\
1 & 0 & 1 & 1 & 2 & 3 & 2 & 3 & 3 \\
2 & 1 & 0 & 2 & 1 & 2 & 3 & 2 & 2 \\
2 & 1 & 2 & 0 & 3 & 4 & 1 & 4 & 4 \\
3 & 2 & 1 & 3 & 0 & 1 & 4 & 1 & 1 \\
4 & 3 & 2 & 4 & 1 & 0 & 5 & 2 & 2 \\
3 & 2 & 3 & 1 & 4 & 5 & 0 & 5 & 5 \\
4 & 3 & 2 & 4 & 1 & 2 & 5 & 0 & 2 \\
4 & 3 & 2 & 4 & 1 & 2 & 5 & 2 & 0
\end{bmatrix}.$$

In the case of a tree, the distance matrix has some attractive properties. As an example, the determinant of the distance matrix of a tree depends only on the number of vertices, and not on the structure of the tree, as seen in the next result.

Theorem 9.2 *Let T be a tree with $V(T) = \{1, \ldots, n\}$. Let D be the distance matrix of T. Then the determinant of D is given by*

$$\det D = (-1)^{n-1}(n-1)2^{n-2}. \tag{9.1}$$

Proof After a relabeling of the vertices we may assume that the vertex n is a pendant and is adjacent to $n-1$. Note that

$$d(i, n) = d(i, n-1) + 1, \quad i = 1, \ldots, n-1.$$

In D, subtract the column $n-1$ from the column n and the row $n-1$ from the row n. Call the resulting matrix D_1. The last row and column of D_1 has all entries 1, except the (n, n)-entry, which is -2. Relabel the vertices $1, \ldots, n-1$, so that $n-1$ is pendant and is adjacent to $n-2$. The resulting matrix is obtained by permuting the rows and the columns of D_1. In that matrix subtract the column $n-2$ from the column $n-1$, and the row $n-2$ from the row $n-1$. Continuing in this way we finally obtain the following matrix

$$D_2 = \begin{bmatrix}
0 & 1 & 1 & \cdots & 1 \\
1 & -2 & 0 & \cdots & 0 \\
1 & 0 & -2 & \cdots & 0 \\
\vdots & \vdots & & \ddots & \vdots \\
1 & 0 & 0 & \cdots & -2
\end{bmatrix}.$$

Clearly $\det D = \det D_2$. By the *Schur complement* formula for the determinant we have,

$$\begin{aligned}
\det D_2 &= \det(-2I_{n-1})(-\mathbf{1}'(0 - 2I_{n-1})^{-1}\mathbf{1}) \\
&= (-2)^{n-1} \times \frac{n-1}{2} \\
&= (-1)^{n-1}(n-1)2^{n-2},
\end{aligned}$$

and the proof is complete.   □

9.1 Distance Matrix of a Graph

We now show that Theorem 9.2 does admit an extension to arbitrary graphs. We need some preliminaries. We assume familiarity with the basic properties of the blocks of a graph. Here we recall the definition and some basic facts. A block of the graph G is a maximal connected subgraph of G that has no cut-vertex. Note that if G is connected and has no cut-vertex, then G itself is a block.

If an edge of a graph is contained in a cycle, then the edge by itself cannot be a block, since it is in a larger subgraph with no cut-vertex. An edge is a block if and only if it is a cut-edge. In particular, the blocks of a tree are precisely the edges of the tree. If a block has more than two vertices, then it is 2-connected. Alternatively, a block of G may be defined as a maximal 2-connected subgraph.

We introduce some notation. If A is an $n \times n$ matrix, then $\operatorname{cof} A$ will denote the sum of all cofactors of A. Note that if A is nonsingular, then

$$\operatorname{cof} A = (\det A)(\mathbf{1}'A^{-1}\mathbf{1}). \tag{9.2}$$

Recall that J denotes the matrix with each entry equal to 1.

Lemma 9.3 *Let A be an $n \times n$ matrix. Then*

$$\det(A + J) = \det A + \operatorname{cof} A. \tag{9.3}$$

Proof First suppose A is nonsingular. By the Schur formula applied in two different ways, we have

$$\det \begin{bmatrix} 1 & -\mathbf{1}' \\ \mathbf{1} & A \end{bmatrix} = \det(A + J) = (\det A)(1 + \mathbf{1}'A^{-1}\mathbf{1}). \tag{9.4}$$

It follows from (9.2) and (9.4) that

$$\det(A + J) = (\det A)\left(1 + \frac{\operatorname{cof} A}{\det A}\right) = \det A + \operatorname{cof} A.$$

When A is singular we may prove the result using a continuity argument, by approximating A by a sequence of nonsingular matrices. $\qquad\square$

As usual, $A(i|j)$ will denote the submatrix obtained by deleting row i and column j of A.

Lemma 9.4 *Let A be an $n \times n$ matrix. Let B be the matrix obtained from A by subtracting the first row from all the other rows and then subtracting the first column from all the other columns. Then*

$$\operatorname{cof} A = \det B(1|1).$$

Proof Let C be the matrix obtained from $A + J$ by subtracting the first row from all the other rows and then subtracting the first column from all the other columns. Let E_{11} be the $n \times n$ matrix with 1 in position $(1, 1)$ and zeros elsewhere. Then $C = B + E_{11}$ and hence $\det C = \det B + \det B(1|1)$. It follows by Lemma 9.3 that

$$\det C = \det(A + J) = \det A + \operatorname{cof} A = \det B + \operatorname{cof} A,$$

and the result follows.   □

Theorem 9.5 *Let G be a connected graph with $V(G) = \{1, \ldots, n\}$. Let $G_1, \ldots, G_k$ be the blocks of G. Then the following assertions hold:*

(i) $\operatorname{cof} D(G) = \displaystyle\prod_{i=1}^{k} \operatorname{cof} D(G_i)$

(ii) $\det D(G) = \displaystyle\sum_{i=1}^{k} \det D(G_i) \prod_{j \neq i} \operatorname{cof} D(G_j).$

Proof We assume, without loss of generality, that G_1 is an end block of G, that is, it contains only one cut-vertex of G. We assume the cut-vertex to be 1. Let $G_1^* = G \setminus (G_1 \setminus \{1\})$ be the remainder of G. Note that the cut-vertex 1 is present in G_1^*. Furthermore, the blocks of G_1^* are $G_2, \ldots, G_k$. We assume $V(G_1) = \{1, \ldots, m\}$ and $V(G_1^*) = \{1, m+1, \ldots, n\}$. Let

$$D(G_1) = \begin{bmatrix} 0 & a' \\ a & E \end{bmatrix}, \quad D(G_1^*) = \begin{bmatrix} 0 & f' \\ f & H \end{bmatrix}.$$

Thus,

$$D(G) = \begin{bmatrix} 0 & a' & f' \\ a & E & a\mathbf{1}' + \mathbf{1}f' \\ f & f\mathbf{1}' + \mathbf{1}a' & H \end{bmatrix}.$$

In $D(G)$ subtract the first column from all the other columns and the first row from all the other rows. The resulting matrix has the same determinant and thus

$$\det D(G) = \det \begin{bmatrix} 0 & a' & f' \\ a & E - a\mathbf{1}' - \mathbf{1}a' & 0 \\ f & 0 & H - f\mathbf{1}' - \mathbf{1}f' \end{bmatrix}$$

$$= \det \begin{bmatrix} 0 & a' \\ a & E - a\mathbf{1}' - \mathbf{1}a' \end{bmatrix} \det(H - f\mathbf{1}' - \mathbf{1}f')$$

$$+ \det \begin{bmatrix} 0 & f' \\ f & H - f\mathbf{1}' - \mathbf{1}f' \end{bmatrix} \det(E - a\mathbf{1}' - \mathbf{1}a'). \qquad (9.5)$$

Clearly,

$$\det D(G_1) = \det \begin{bmatrix} 0 & a' \\ a & E \end{bmatrix} = \det \begin{bmatrix} 0 & a' \\ a & E - a\mathbf{1}' - \mathbf{1}a' \end{bmatrix} \qquad (9.6)$$

and

$$\det D(G_1^*) = \det \begin{bmatrix} 0 & f' \\ f & H \end{bmatrix} = \det \begin{bmatrix} 0 & f' \\ f & H - f\mathbf{1}' - \mathbf{1}f' \end{bmatrix}. \qquad (9.7)$$

It follows from Lemma 9.4 that

$$\mathrm{cof}\, D(G_1) = \det(E - a\mathbf{1}' - \mathbf{1}a') \qquad (9.8)$$

and

$$\mathrm{cof}\, D(G_1^*) = \det(H - f\mathbf{1}' - \mathbf{1}f'). \qquad (9.9)$$

Substituting (9.6), (9.7), (9.8) and (9.9) in (9.5) we get

$$\det D(G) = \det D(G_1)\mathrm{cof}\, D(G_1^*) + \det D(G_1^*)\mathrm{cof}\, D(G_1) \qquad (9.10)$$

It also follows from Lemma 9.4 that

$$\begin{aligned} \mathrm{cof}\, D(G) &= \det \begin{bmatrix} E - a\mathbf{1}' - \mathbf{1}a' & 0 \\ 0 & H - f\mathbf{1}' - \mathbf{1}f' \end{bmatrix} \\ &= \det(E - a\mathbf{1}' - \mathbf{1}a')\det(H - f\mathbf{1}' - \mathbf{1}f') \\ &= \mathrm{cof}\, D(G_1)\mathrm{cof}\, D(G_1^*) \end{aligned} \qquad (9.11)$$

Note that the proof is complete at this point if $k = 2$, while the result is trivial if G is itself a block. We prove the result by induction on k. Since G_1^* has blocks $G_2, \ldots, G_k$, by induction assumption we have

$$\mathrm{cof}\, D(G_1^*) = \prod_{i=2}^{k} \mathrm{cof}\, D(G_i) \qquad (9.12)$$

and

$$\det D(G_1^*) = \sum_{i=2}^{k} \det D(G_i) \prod_{j \neq i} \mathrm{cof}\, D(G_j). \qquad (9.13)$$

The proof is completed by substituting (9.12) and (9.13) in (9.10) and (9.11). $\square$

According to Theorem 9.5 the determinant of the distance matrix of a graph does not change if the blocks of the graph are reassembled in some other way. In the case of a tree the blocks are precisely the edges, and thus the determinant of the distance matrix of a tree depends only on the number of edges. The formula given in Theorem 9.2 follows easily from Theorem 9.5. To see this suppose T is a tree with

$V(T) = \{1, \ldots, n\}$ and $E(T) = \{e_1, \ldots, e_{n-1}\}$. Then the blocks of T are the edges; more precisely, the blocks $G_1, \ldots, G_{n-1}$ are the graphs induced by $e_1, \ldots, e_{n-1}$, respectively. Then $D(G_i) = \begin{bmatrix} 0 & 1 \\ 1 & 0 \end{bmatrix}$, $i = 1, \ldots, n-1$, and hence $\det D(G_i) = -1$ and $\operatorname{cof} D(G_i) = -2$, $i = 1, \ldots, n-1$. It follows by (ii) of Theorem 9.5 that

$$\det D(T) = (n-1)(-1)^{n-1}2^{n-2},$$

which is Theorem 9.2.

In the case of a unicyclic graph, Theorem 9.5 implies that the determinant of the distance matrix depends only on the length of the cycle and the number of edges.

9.2 Distance Matrix and Laplacian of a Tree

Let T be a tree with $V(T) = \{1, \ldots, n\}$. Let D be the distance matrix of T and L the Laplacian of T. It follows by Theorem 9.2 that D is nonsingular. It is an interesting and unexpected fact that the inverse of D is related to the Laplacian through a rather simple formula. Before proving the formula we need some preliminaries. As usual, let d_i be the degree of the vertex i and let $\tau_i = 2 - d_i$, $i = 1, \ldots, n$. Let τ be the $n \times 1$ vector with components $\tau_1, \ldots, \tau_n$. We note an easy property of τ. Recall that $\mathbf{1}$ denotes the vector of 1 s of appropriate size.

Lemma 9.6 $\mathbf{1}'\tau = 2$.

Proof Since the sum of the degrees of all the vertices is twice the number of edges, we have

$$\sum_{i=1}^{n} \tau_i = \sum_{i=1}^{n} (2 - d_i) = 2n - 2(n-1) = 2.$$

This completes the proof.   □

Lemma 9.7 *Let T be a tree with $V(T) = \{1, \ldots, n\}$. Let D be the distance matrix of T. Then*

$$D\tau = (n-1)\mathbf{1}. \tag{9.14}$$

Proof We prove the result by induction on n. The result is obvious for $n = 1$. For $n = 2$, $D = \begin{bmatrix} 0 & 1 \\ 1 & 0 \end{bmatrix}$ and $\tau = \begin{bmatrix} 1 \\ 1 \end{bmatrix}$. Then it is easily verified that $D\tau = \mathbf{1}$. So let $n \geq 3$ and assume the result to be true for trees with less than n vertices. We may assume, without loss of generality, that the vertex n is pendant and is adjacent to $n - 1$. Partition D and τ as

$$D = \begin{bmatrix} D(n, n) & x \\ x' & 0 \end{bmatrix}, \quad \tau = \begin{bmatrix} \tau(n) \\ 1 \end{bmatrix}.$$

Note that $\tau_n = 2 - d_n = 1$. Then

$$D\tau = \begin{bmatrix} D(n, n)\tau(n) + x \\ x'\tau(n) \end{bmatrix}.$$

$$(9.15)$$

The distance matrix of $T \setminus \{n\}$ is $D(n, n)$. Furthermore, the degree of the vertex $n - 1$ in $T \setminus \{n\}$ is $d_{n-1} - 1$. Let $\hat{\tau}$ be the vector obtained by adding 1 to the last component of $\tau(n)$. By an induction assumption,

$$D(n, n)\hat{\tau} = (n - 2)\mathbf{1}.$$

$$(9.16)$$

Let y be the last column of $D(n, n)$. It follows from (9.16) that

$$D(n, n)\tau(n) = (n - 2)\mathbf{1} - y.$$

$$(9.17)$$

Since $d(i, n) = d(i, n - 1) + 1$, $i = 1, 2, \ldots, n - 1$, then

$$x = y + \mathbf{1}.$$

$$(9.18)$$

It follows from (9.17) and (9.18) that

$$D(n, n)\tau(n) + x = (n - 2)\mathbf{1} - x + \mathbf{1} + x = (n - 1)\mathbf{1}.$$

$$(9.19)$$

It follows from (9.15) and (9.19) that (9.14) is valid, except possibly in coordinate n, which corresponds to a pendant vertex. Since a tree with 3 or more vertices has at least 2 pendant vertices, we may repeat the argument with another pendant vertex and conclude that (9.14) holds in the coordinate n as well. This completes the proof.

$\square$

Lemma 9.8 *Let T be a tree with $V(T) = \{1, \ldots, n\}$. Let D be the distance matrix of T and let L be the Laplacian of T. Then*

$$LD + 2I = \tau \mathbf{1}'.$$

$$(9.20)$$

Proof Fix vertices $i, j \in \{1, \ldots, n\}$. Let the degree of i be $d_i = k$, and we assume, without loss of generality, that i is adjacent to $\{1, \ldots, k\}$. First suppose $i \neq j$. The graph $T \setminus \{i\}$ is a forest with k components and we assume, without loss of generality, that j is in the component of $T \setminus \{i\}$ that contains 1. Then

$$d(v, j) = d(1, j) + 2 = d(i, j) + 1, \quad v = 2, \ldots, k.$$

$$(9.21)$$

It follows from (9.21) that

$$
\begin{aligned}
(LD + 2I)_{ij} &= (LD)_{ij} \\
&= d_i d(i, j) - (d(1, j) + \cdots + d(k, j)) \\
&= kd(i, j) - (kd(i, j) + k - 2) \\
&= 2 - k \\
&= \tau_i.
\end{aligned}
\tag{9.22}
$$

If $j = i$ then

$$
\begin{aligned}
(LD + 2I)_{ii} &= -(d(i, 1) + \cdots + d(i, k)) + 2 \\
&= 2 - k \\
&= \tau_i.
\end{aligned}
\tag{9.23}
$$

It follows from (9.22) and (9.23) that

$$
(LD + 2I)_{ij} = \tau_i
$$

for all i, j and hence (9.20) holds. This completes the proof. $\square$

We are now in a position to give a formula for the inverse of the distance matrix in terms of the Laplacian.

Theorem 9.9 *Let T be a tree with $V(T) = \{1, \ldots, n\}$. Let D be the distance matrix of T and L be the Laplacian of T. Then*

$$
D^{-1} = -\frac{1}{2}L + \frac{1}{2(n-1)}\tau\tau'.
\tag{9.24}
$$

Proof We have

$$
\begin{aligned}
\left(-\frac{1}{2}L + \frac{1}{2(n-1)}\tau\tau'\right) D &= -\frac{1}{2}LD + \frac{1}{2(n-1)}\tau\tau'D \\
&= -\frac{1}{2}LD + \frac{1}{2}\tau\mathbf{1}' && \text{by Lemma 9.7} \\
&= -\frac{1}{2}(LD - \tau\mathbf{1}') \\
&= -\frac{1}{2}(-2I) && \text{by Lemma 9.8} \\
&= I.
\end{aligned}
$$

Therefore (9.24) holds and the proof is complete. $\square$

We introduce some notation. Let $i, j \in \{1, \ldots, n\}$, $i \neq j$. Denote by e_{ij} the $n \times 1$ vector with the ith coordinate equal to 1, the jth coordinate equal to -1, and zeros elsewhere. Note that if B is an $n \times n$ matrix then

$$e'_{ij} B e_{ij} = b_{ii} + b_{jj} - b_{ij} - b_{ji}.$$

Recall that H is a g-inverse of A if $AHA = A$.

Lemma 9.10 *Let G be a connected graph with $V(G) = \{1, \ldots, n\}$, and L be the Laplacian of G. Let $i, j \in \{1, \ldots, n\}$, $i \neq j$. If H^1 and H^2 are any two g-inverses of L, then*

$$e'_{ij} H^1 e_{ij} = e'_{ij} H^2 e_{ij}.$$

Proof Since G is connected, by Lemma 4.3 the rank of L is $n - 1$. Thus, $\mathbf{1}$ is the only vector in the null space of L, up to a scalar multiple. Since $\mathbf{1}' e_{ij} = 0$, then e_{ij} is in the column space of L. Therefore, there exists a vector z such that $e_{ij} = Lz$. Then

$$e'_{ij}(H^1 - H^2)e_{ij} = z'L(H^1 - H^2)Lz = z'(LH^1L - LH^2L)z = 0,$$

since $LH^1L = LH^2L = L$. This completes the proof. $\qquad\square$

Lemma 9.11 *Let T be a tree with $V(T) = \{1, \ldots, n\}$. Let D be the distance matrix and L, the Laplacian of T. Then*

$$LDL = -2L.$$

Proof By Lemma 9.8,

$$LD + 2I = \tau \mathbf{1}'. \tag{9.25}$$

Post-multiplying (9.25) by L, and keeping in view that $L\mathbf{1} = 0$, we have

$$LDL + 2L = \tau \mathbf{1}' L = 0,$$

and the proof is complete. $\qquad\square$

Recall that the Moore–Penrose inverse of the matrix B is the unique g-inverse B^+ of B that satisfies $B^+ B B^+ = B^+$ and that BB^+ and $B^+ B$ are symmetric.

Theorem 9.12 *Let T be a tree with $V(T) = \{1, \ldots, n\}$. Let D be the distance matrix and L the Laplacian T. If H is a g-inverse of L then*

$$h_{ii} + h_{jj} - h_{ij} - h_{ji} = d(i, j).$$

In particular,

$$d(i, j) = \ell_{ii}^+ + \ell_{jj}^+ - 2\ell_{ij}^+. \tag{9.26}$$

Proof Let $S = -\frac{D}{2}$. By Lemma 9.11, S is a g-inverse of L. It follows by Lemma 9.10 that

$$h_{ii} + h_{jj} - h_{ij} - h_{ji} = s_{ii} + s_{jj} - s_{ij} - s_{ji}.$$

The result follows in view of $s_{ii} = s_{jj} = 0$ and $s_{ij} = s_{ji} = -\frac{d(i,j)}{2}$. $\square$

Let G be a graph with $V(G) = \{1, \ldots, n\}$. Let $D = [d(i,j)]$ be the distance matrix of G. The Wiener index $W(G)$ of G, which has applications in mathematical chemistry, is defined as

$$W(G) = \sum_{i<j} d(i,j).$$

Lemma 9.13 *Let T be a tree with $V(T) = \{1, \ldots, n\}$. Let D be the distance matrix and L, the Laplacian of T. Let $\lambda_1 \geq \cdots \geq \lambda_{n-1} > \lambda_n = 0$ be the eigenvalues of L. Then*

$$W(T) = n \sum_{i=1}^{n-1} \frac{1}{\lambda_i}.$$

Proof Note that $L^+\mathbf{1} = L^+LL^+\mathbf{1} = (L^+)^2L\mathbf{1} = 0$, that is, the row sums of L^+ are zero. By (9.26),

$$d(i,j) = \ell_{ii}^+ + \ell_{jj}^+ - 2\ell_{ij}^+. \tag{9.27}$$

Summing both sides of (9.27) with respect to i, j and using $L^+\mathbf{1} = 0$, we get

$$\sum_{i<j} d(i,j) = \frac{1}{2} \sum_{i=1}^{n} \sum_{j=1}^{n} d(i,j) = n \sum_{i=1}^{n} \ell_{ii}^+. \tag{9.28}$$

The eigenvalues of L^+ are given by $\frac{1}{\lambda_{n-1}} \geq \cdots \geq \frac{1}{\lambda_1} > 0$. It follows from (9.28) that

$$W(G) = \sum_{i<j} d(i,j) = n \sum_{i=1}^{n} \ell_{ii}^+ = n(\text{trace } L^+) = n \sum_{i=1}^{n} \frac{1}{\lambda_i}$$

and the proof is complete. $\square$

Let T be a tree with $V(T) = \{1, \ldots, n\}$. Let D be the distance matrix and L, the Laplacian of T. Suppose each edge of T is oriented and let Q be the $n \times (n-1)$ vertex-edge incidence matrix of T. Then $L = QQ'$. With this notation we have the following result.

Lemma 9.14 $Q'DQ = -2I$.

Proof By Lemma 9.11, $LDL = -2L$ and hence

$$QQ'DQQ' = -2QQ'. \tag{9.29}$$

By Lemma 2.2, Q has full column rank and hence it admits a left inverse, say H. It follows from (9.29) that

$$HQQ'DQQ'H' = -2HQQ'H'$$

and hence $Q'DQ = -2I$. This completes the proof.   □

9.3 Eigenvalues of the Distance Matrix of a Tree

We begin with an observation, which is an immediate consequence of Theorem 9.2.

Lemma 9.15 Let T be a tree with $V(T) = \{1, \ldots, n\}$, $n \geq 2$. Let D be the distance matrix of T. Then D has 1 positive and $n - 1$ negative eigenvalues.

Proof If $n = 2$ then $D = \begin{bmatrix} 0 & 1 \\ 1 & 0 \end{bmatrix}$, which has eigenvalues 1 and -1. Assume that the result is true for a tree with $n - 1$ vertices and proceed by induction on n. If vertex i is a pendant vertex of T, then the matrix $D(i, i)$, obtained by deleting row and column i of D, is the distance matrix of the tree $T \setminus \{i\}$. By an induction assumption, $D(i, i)$ has 1 positive and $n - 2$ negative eigenvalues. It follows by the interlacing theorem that D has either 1 or 2 positive eigenvalues. By Theorem 9.2,

$$\frac{\det D}{\det D(i, i)} = -\frac{2(n - 1)}{n - 2} < 0.$$

Thus, D must have 1 positive eigenvalue.   □

We now obtain an interlacing inequality connecting the eigenvalues of the distance matrix and the Laplacian.

Theorem 9.16 Let T be a tree with $V(T) = \{1, \ldots, n\}$. Let D be the distance matrix and L the Laplacian of T. Let $\mu_1 > 0 > \mu_2 \geq \cdots \geq \mu_n$ be the eigenvalues of D and let $\lambda_1 \geq \cdots \geq \lambda_{n-1} > \lambda_n = 0$ be the eigenvalues of L. Then

$$0 > -\frac{2}{\lambda_1} \geq \mu_2 \geq -\frac{2}{\lambda_2} \geq \cdots \geq -\frac{2}{\lambda_{n-1}} \geq \mu_n.$$

Proof Let each edge of T be given an orientation and let Q be the $n \times (n-1)$ vertex-edge incidence matrix. There exists an $(n - 1) \times (n - 1)$ nonsingular matrix M such that the columns of QM are orthonormal. (This follows from an application of the Gram–Schmidt process on the columns of Q.) Since $1'Q = 0$ it is easily verified that the matrix U defined as $U = \begin{bmatrix} QM, & \frac{1}{\sqrt{n}}1 \end{bmatrix}$ is orthogonal. Now

$$U'DU = \begin{bmatrix} M'Q'DQM & \frac{1}{\sqrt{n}}M'Q'D1 \\ \frac{1}{\sqrt{n}}1'DQM & \frac{1}{n}1'D1 \end{bmatrix}. \tag{9.30}$$

Let $K = Q'Q$ be the edge-Laplacian matrix. Then K is nonsingular and $M'KM = M'Q'QM = I$. Hence, $K^{-1} = MM'$. Thus, K^{-1} and $M'M$ have the same eigenvalues. It follows from Lemma 9.14 and (9.30) that the leading $(n-1) \times (n-1)$ principal submatrix of $U'DU$ is $-2M'M$. By the interlacing theorem the eigenvalues of $U'DU$, which are the same as the eigenvalues of D, interlace the eigenvalues of $-2M'M = -2K^{-1}$. The eigenvalues of K are the same as the nonzero eigenvalues of L. Hence the eigenvalues of K^{-1} are the same as the nonzero eigenvalues of L^+, and the proof is complete. $\square$

We now obtain some results for the eigenvalues of the Laplacian of a tree. We may then use Theorem 9.16 to obtain results for the eigenvalues of the distance matrix.

Theorem 9.17 *Let T be a tree with $V(T) = \{1, \ldots, n\}$. Let L be the Laplacian of T. Suppose $\mu > 1$ is an integer eigenvalue of L with u as a corresponding eigenvector. Then the following assertions hold:*

(i) *μ divides n.*
(ii) *No coordinate of u is zero.*
(iii) *The algebraic multiplicity of μ is 1.*

Proof Since $\det L = 0$ then zero is an eigenvalue of L, and hence the characteristic polynomial $\det(\lambda I - L)$ of L is of the form $\lambda f(\lambda)$. We may write $f(\lambda) = \lambda g(\lambda) - f(0)$, where $g(\lambda)$ is a polynomial with integer coefficients. The coefficient of λ in the characteristic polynomial of L is, up to a sign, the sum of the $(n-1) \times (n-1)$ principal minors of L, which equals n, since by the matrix-tree theorem, each cofactor of L is 1. It follows that $f(0) = n$. Thus, $0 = f(\mu) = \mu g(\mu) - n$ and hence $\mu g(\mu) = n$. This proves (i).

To prove (ii) suppose u has a zero coordinate and, without loss of generality, let $u_n = 0$. Let $T_1, \ldots, T_k$ be the components of $T \setminus \{n\}$. Partition L and u conformally so that $Lu = \mu u$ is expressed as

$$\begin{bmatrix} L_1 & 0 & \cdots & 0 \\ 0 & L_2 & \cdots & 0 \\ \vdots & \vdots & \ddots & \vdots & w \\ 0 & 0 & \cdots & L_k \\ & & w' & & d_n \end{bmatrix} \begin{bmatrix} u^1 \\ u^2 \\ \vdots \\ u^k \\ 0 \end{bmatrix} = \mu \begin{bmatrix} u^1 \\ u^2 \\ \vdots \\ u^k \\ 0 \end{bmatrix}, \qquad (9.31)$$

where L_j is the submatrix of L corresponding to vertices of $T_j, j = 1, \ldots, k$.

Since $u^i \neq 0$ for some $i = 1, \ldots, k$, we assume, without loss of generality, that $u^1 \neq 0$. Then $L_1 u^1 = \mu u^1$ implies that μ is an eigenvalue of L_1. There must be a vertex of T_1 which is adjacent to n and, without loss of generality, we assume that 1 is adjacent to n. Then $L_1 = L(T_1) + E_{11}$, where $L(T_1)$ is the Laplacian of T_1 and E_{11} is the matrix with 1 at position $(1, 1)$ and zeros elsewhere. Then $\det(L_1)$ can be seen to be equal to $\det L(T_1)$, which is zero plus the cofactor of the $(1, 1)$, element of $L(T_1)$, which is 1 by the matrix-tree theorem. Thus, $\det(L_1) = 1$. It follows that L_1^{-1} is an integer matrix. Since any rational root of a polynomial with integer coefficients must be an integer, any rational eigenvalue of L_1^{-1} must be an integer. However, $\frac{1}{\mu}$ is an eigenvalue of L_1^{-1}, which is a contradiction. This proves (ii).

If there are two linearly independent eigenvectors of L corresponding to μ, we can produce an eigenvector with a zero coordinate, contradicting (ii). Hence the algebraic multiplicity of μ is 1 and the proof is complete. □

We introduce some notation. For a tree T, $p(T)$ will denote the number of pendant vertices of T. A vertex is called *quasipendant* if it is adjacent to a pendant vertex. The number of quasipendant vertices will be denoted by $q(T)$. We assume that each edge of the tree is oriented, let Q be the vertex-edge incidence matrix and $K = Q'Q$ the edge-Laplacian.

Theorem 9.18 *Let T be a tree with $V(T) = \{1, \dots, n\}, n \geq 2$, and L be the Laplacian of T. If μ is an eigenvalue of L then the algebraic multiplicity of μ is at most $p(T) - 1$.*

Proof Let $k = p(T)$. We assume, without loss of generality, that $1, \dots, k$ are the pendant vertices of T, and furthermore, 1 is adjacent to the quasipendant vertex $k+1$. We first make the following claim, which we will prove by induction on n. The claim is that if x is an eigenvector of L, then among $x_1, \dots, x_k$, at least two coordinates must be nonzero. To prove the claim let x be an eigenvector of L corresponding to μ. If $x_1, \dots, x_k$ are all nonzero, there is nothing to prove. So suppose $x_1 = 0$. Let y be the vector obtained by deleting x_1 from x. (We continue to list the coordinates of y as $y_2, \dots, y_n$ rather than as $y_1, \dots, y_{n-1}$.) From $Lx = \mu x$ it follows that $x_{k+1} = 0$, and that y is an eigenvector of the Laplacian of $T \setminus \{1\}$ for μ. The pendant vertices of $T \setminus \{1\}$ are either $\{2, \dots, k\}$ or $\{2, \dots, k+1\}$. Since $y_{k+1} = x_{k+1} = 0$, by an induction assumption it follows that at least two coordinates among $y_2, \dots, y_k$ must be nonzero, and the claim is proved.

Suppose the multiplicity of μ is at least $p(T) = k$. Let $z^1, \dots, z^k$ be linearly independent eigenvectors of L corresponding to μ. We may find a linear combination z of $z^1, \dots, z^k$ such that among the first k coordinates of z, at most one is nonzero. Then z is an eigenvector of L for which the claim proved earlier does not hold. This contradiction proves that the multiplicity of μ is at most $k - 1$. □

Corollary 9.19 *Let T be a tree with $V(T) = \{1, \dots, n\}, n \geq 2$. Let D be the distance matrix of T. If μ is an eigenvalue of D then the algebraic multiplicity of μ is at most $p(T)$.*

Proof If the algebraic multiplicity of μ is greater than $p(T)$, then by Theorem 9.16 the multiplicity of $-\frac{2}{\mu}$, as an eigenvalue of $-2K^{-1}$, will be greater than $p(T) - 1$. But then the multiplicity of $-\frac{\mu}{2}$, as an eigenvalue of L, will be greater than $p(T) - 1$, contradicting Theorem 9.18. □

Theorem 9.20 *Let T be a tree with $V(T) = \{1, \dots, n\}, n \geq 2$. Let L be the Laplacian of T. Then $\mu = 1$ is an eigenvalue of L with multiplicity at least $p(T) - q(T)$.*

Proof Let $s = q(T)$, let $1, \ldots, s$ be the quasipendant vertices of T and suppose they are adjacent to $r_1, \ldots, r_s$ pendant vertices, respectively. (These pendant vertices are necessarily distinct, since the same pendant vertex cannot be adjacent to two quasipendant vertices.) Recall that for vertices $i \neq j$, e_{ij} is the $n \times 1$ vector with 1 at the ith place, -1 at the jth place, and zeros elsewhere. Suppose i and j are pendant vertices of T, adjacent to a common quasipendant vertex. Then it is easily verified that e_{ij} is an eigenvector of L, corresponding to the eigenvalue 1. This way we can generate $(r_1 - 1) + \cdots + (r_s - 1)$ linearly independent eigenvectors of L corresponding to the pendant vertices for the eigenvalue 1. Hence the multiplicity of 1, as an eigenvalue of L, is at least

$$\sum_{i=1}^{s}(r_i - 1) = \sum_{i=1}^{s} r_i - s = p(T) - q(T)$$

and the proof is complete.   □

Corollary 9.21 *Let T be a tree with $V(T) = \{1, \ldots, n\}$, $n \geq 2$. Let D be the distance matrix of T. Then -2 is an eigenvalue of D with multiplicity at least $p(T) - q(T) - 1$.*

Proof The result follows from Theorem 9.20 and Theorem 9.16.   □

Example 9.22 Consider the tree T:

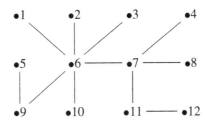

Then T has eight pendant vertices: $1, 2, 3, 4, 5, 8, 10, 12$, and four quasipendant vertices: $6, 7, 9, 11$. Thus, $p(T) = 8$ and $q(T) = 4$. Therefore, the Laplacian of T has 1 as an eigenvalue with multiplicity at least 4, while the distance matrix of T has -2 as an eigenvalue with multipliity at least 3. In this case the actual multiplicities can be verified to be 4 in both the cases.

Exercises

1. Let T be a tree with $V(T) = \{1, \ldots, n\}$ and $E(T) = \{e_1, \ldots, e_{n-1}\}$. Suppose edge e_i is given the weight w_i, $i = 1, \ldots, n - 1$. The distance between vertices i and j is defined to be the sum of the weights of the edges on the unique ij-path. The distance matrix D is the $n \times n$ matrix with d_{ij} equal to the distance between

i and j if $i \neq j$, and $d_{ii} = 0$, $i = 1, \ldots, n$. Show that

$$\det D = (-1)^{n-1} 2^{n-2} \left(\sum_{i=1}^{n-1} w_i \right) \prod_{i=1}^{n-1} w_i.$$

2. Let T be a tree with $V(T) = \{1, \ldots, n\}$ and $E(T) = \{e_1, \ldots, e_{n-1}\}$. For a real number q, the q-distance matrix $D_q = [d_{ij}^q]$ of T is the $n \times n$ matrix defined as follows: d_{ij}^q is equal to $1 + q + q^2 + \cdots + q^{d(i,j)-1}$ if $i \neq j$, and $d_{ii}^q = 0$, $i = 1, \ldots, n$. Show that

$$\det D_q = (-1)^{n-1} (n-1)(1+q)^{n-2}.$$

3. Let T be a tree with $V(T) = \{1, \ldots, n\}$. Let D be the distance matrix of T. As usual, let J be the matrix of all 1 s. Show that for any real number α,

$$\det(D + \alpha J) = (-1)^{n-1} 2^{n-2} (2\alpha + n - 1).$$

4. Let T be a tree with $V(T) = \{1, \ldots, n\}$. Let D be the distance matrix and L the Laplacian of T. Show that $(D^{-1} - L)^{-1} = \frac{1}{3}(D + (n-1)J)$.

5. Let T be a tree and let G be a graph with $V(T) = V(G) = \{1, \ldots, n\}$. Let D be the distance matrix of T and let S be the Laplacian of G. Show that $D^{-1} - S$ is nonsingular.

6. Let T be a tree with $V(T) = \{1, \ldots, n\}$, $n \geq 2$. Let D be the distance matrix and L the Laplacian of T. Fix $i, j \in \{1, \ldots, n\}$, $i \neq j$. Define the $n \times n$ matrix H as follows. The ith row and column of H has all zeros, while $H(i, i) = L(i, i)^{-1}$. Show that H is a g-inverse of L. Hence, using the fact that $e_{ij}' H e_{ij} = d(i, j)$, conclude that $d(i, j) = \det L(i, j; i, j)$, where $L(i, j; i, j)$ is the submatrix of L obtained by deleting rows i, j and columns i, j. (This provides another proof of Corollary 4.10.)

7. Let T be a tree with $V(T) = \{1, \ldots, n\}$, and suppose n is odd. Show that the Wiener index of T is even.

8. Let T be a tree with $V(T) = \{1, \ldots, n\}$. Let D be the distance matrix of T. Show that D is a squared Euclidean distance matrix, that is, there exist points $x^1, \ldots, x^n$ in $\mathbb{R}^k$ for some k such that $d(i, j) = ||x^i - x^j||^2$, $i, j = 1, \ldots, n$.

9. Let T be a tree with $V(T) = \{1, \ldots, n\}$. Let $i, j, k, \ell \in \{1, \ldots, n\}$ be four vertices of T, which are not necessarily distinct. Show that among the three numbers $d(i, j) + d(k, \ell)$, $d(i, k) + d(j, \ell)$ and $d(i, \ell) + d(j, k)$, two are equal and are not less than the third.

10. Let T be a tree with $V(T) = \{1, \ldots, n\}$. Let D be the distance matrix of T and let $\mu_1 > 0 > \mu_2 \geq \cdots \geq \mu_n$ be the eigenvalues of D. Suppose T has k pendant vertices. Show that $\mu_k \geq -2$ and $\mu_{n-k+2} \leq -2$.

Theorem 9.2 is due to Graham and Pollak [GP71]. Section 8.1 is based on [GHH77].
An extension of Theorem 9.5 for a more general class of "additive distances", which
includes resistance distance, has been given in [BG08]. Theorem 9.9 is due to Graham
and Lovász [Gra78]. The proof technique and several other results in Sect. 8.2 are
adapted from [Bap04, BKN05]. Section 8.3 is based on [Far85, GMS90, Mer90].
Exercises 1–5 are based on [BKN05, BLP06].

References and Further Reading

[Bap04]   Bapat, R.B.: Resistance matrix of a weighted graph. MATCH Commun. Math. Comput.
           Chem. **50**, 73–82 (2004)
[BG08]   Bapat, R.B., Gupta, S.: Resistance matrices of blocks in a graph. AKCE Int. J. Graphs
           Comb. **5**(1), 35–45 (2008)
[BKN05]   Bapat, R., Kirkland, S.J., Neumann, M.: On distance matrices and Laplacians. Linear
           Algebra Appl. **401**, 193–209 (2005)
[BLP06]   Bapat, R.B., Lal, A.K., Pati, S.: A q-analogue of the distance matrix of a tree. Linear
           Algebra Appl. **416**, 799–814 (2006)
[Far85]   Faria, I.: Permanental roots and the star degree of a graph. Linear Algebra Appl. **64**,
           255–265 (1985)
[GHH77]   Graham, R.L., Hoffman, A.J., Hosoya, H.: On the distance matrix of a directed graph.
           J. Comb. Theor. **1**, 85–88 (1977)
[Gra78]   Graham, R.L., Lovász, L.: Distance matrix polynomials of trees. Adv. in Math. **29**(1),
           60–88 (1978)
[GP71]   Graham, R.L., Pollak, H.O.: On the addressing problem for loop switching. Bell. System
           Tech. J. **50**, 2495–2519 (1971)
[GMS90]   Grone, R., Merris, R., Sunder, V.S.: The Laplacian spectrum of a graph. SIAM J. Matrix
           Anal. Appl. **11**, 218–238 (1990)
[Mer90]   Merris, R.: The distance spectrum of a tree. J. Graph Theor. **3**(14), 365–369 (1990)

Chapter 10
Resistance Distance

Let G be a connected graph with $V(G) = \{1, \ldots, n\}$. The shortest path distance $d(i, j)$ between the vertices $i, j \in V(G)$ is the classical notion of distance and is extensively studied. However, this concept of distance is not always appropriate. Consider the following two graphs:

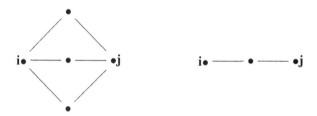

In both theses graphs, $d(i, j) = 2$. But it is evident that in the first graph there are more paths connecting i and j (we might say that there is a better "communication" between i and j), and hence it is reasonable that the "distance" between i and j should be smaller in the first graph in comparison to that in the second graph. This feature is not captured by classical distance. Also, classical distance is not mathematically tractable unless the graph is a tree.

Another notion of distance, called "resistance distance", in view of an interpretation of the notion vís-a-vís resistance in electrical networks, captures the notion of distance in terms of communication more appropriately. Resistance distance is mathematically more tractable, as well. Furthermore, in the case of a tree, resistance distance and classical distance coincide.

Resistance distance admits several equivalent definitions. As a starting point we present a definition in terms of g-inverse.

Let G be a connected graph with $V(G) = \{1, \ldots, n\}$ and let L be the Laplacian of G. We assume that the edges of G are oriented, although the orientations do not play any role as far as the resistance distance is concerned. Let $i, j \in \{1, \ldots, n\}$, $i \neq j$. Recall the definition of the $n \times 1$ vector e_{ij}, which has a 1 at the ith place, a -1 at the jth place, and zeros elsewhere. By Lemma 9.10, $e_{ij}' H e_{ij}$ is invariant with respect to a

© Springer-Verlag London 2014
R.B. Bapat, *Graphs and Matrices*, Universitext,
DOI 10.1007/978-1-4471-6569-9_10

g-inverse H of L. We define the resistance distance between i and j, denoted $r(i, j)$, as

$$r(i, j) = e'_{ij} H e_{ij} = h_{ii} + h_{jj} - h_{ij} - h_{ji},$$

where H is a g-inverse of L. If $i = j$ then we set $r(i, j) = 0$.

We remark that if H is a symmetric g-inverse of L then $r(i, j) = h_{ii} + h_{jj} - 2h_{ij}$. In particular, setting $M = L^+$, $r(i, j) = m_{ii} + m_{jj} - 2m_{ij}$.

If G is disconnected then we may define the resistance distance between the two vertices i and j in the same component of G if we restrict ourselves to that component. The resistance distance between vertices in different components may be defined as infinity, although we will not encounter this case.

10.1 The Triangle Inequality

Let G be a connected graph with $V(G) = \{1, \ldots, n\}$ and let $\rho : V(G) \times V(G) \to \mathbb{R}$. If ρ is to represent a measure of distance between a pair of vertices then it is reasonable to expect that ρ should satisfy the following properties:

(i)   (Nonnegativity) $\rho(i, j) \geq 0$ for all i, j, with equality if and only if $i = j$.
(ii) (Symmetry) $\rho(i, j) = \rho(j, i)$.
(iii) (Triangle inequality) $\rho(i, j) + \rho(j, k) \geq \rho(i, k)$.

Classical distance $d(i, j)$ clearly satisfies these properties. We now show that these properties are enjoyed by resistance distance as well.

If $n \leq 2$, then the properties are easy to prove, so assume $n \geq 3$. Let L be the Laplacian matrix of G and let $M = L^+$. Since L is symmetric, so is M. Also, L is positive semidefinite and it follows that $M = MLM$ is also positive semidefinite. Thus,

$$r(i, j) = e'_{ij} M e_{ij} \geq 0.$$

Since $LML = L$ and $MLM = M$, then rank M = rank L, and as noted in Lemma 4.3 rank $L = n - 1$ since G is connected. Thus, any 2×2 principal minor of M is positive, i.e., for $i \neq j$, $m_{ii} m_{jj} > m_{ij}^2$. It follows by the arithmetic mean–geometric mean inequality that $m_{ii} + m_{jj} > 2m_{ij}$. Thus, for any $i \neq j$,

$$r(i, j) = m_{ii} + m_{jj} - 2m_{ij} > 0.$$

This shows that the resistance distance satisfies (i). Clearly,

$$r(i, j) = e'_{ij} M e_{ij} = r(j, i)$$

and hence (ii) holds. We now show that the resistance distance satisfies (iii). We first prove a preliminary result.

Lemma 10.1 *Let G be a connected graph with n vertices and let L be the Laplacian of G. If B is any proper principal submatrix of L, then B^{-1} is an entrywise nonnegative matrix.*

Proof Let B be a $k \times k$ principal submatrix of L, $1 \le k \le n-1$. Since det $B > 0$, B is nonsingular. We prove the result by induction on k. The proof is easy for $k \le 2$. Assume the result to be true for principal submatrices of order less than k. It will be sufficient to show that all the cofactors of B are nonnegative. The cofactor of a diagonal entry of B is the determinant of a principal submatrix of L, which is positive. We show that the cofactor of the $(1, 2)$-element of B is nonnegative, and the case of other cofactors will be similar. Partition $B(1|2)$ as

$$B(1|2) = \begin{bmatrix} b_{21} & x' \\ y & B(1, 2|1, 2) \end{bmatrix}.$$

Then

$$\det B(1|2) = (\det B(1, 2|1, 2))(b_{21} - x'(B(1, 2|1, 2)^{-1}y). \tag{10.1}$$

By induction assumption, $B(1, 2|1, 2)^{-1} \ge 0$. Also x and y have all entries nonpositive. Furthermore, det $B(1, 2|1, 2) > 0$ and $b_{21} \le 0$. It follows from (10.1) that det $B(1|2) \le 0$. Thus, the cofactor of the $(1, 2)$-element of B is nonnegative and the proof is complete. □

We return to the proof of the fact that the resistance distance satisfies the triangle inequality. In order to prove $r(i, j) + r(j, k) \ge r(i, k)$, we must show that for any g-inverse H of L,

$$h_{ii} + h_{jj} - 2h_{ij} + h_{jj} + h_{kk} - 2h_{jk} \ge h_{ii} + h_{kk} - 2h_{ik},$$

and this is equivalent to

$$h_{jj} + h_{ik} - h_{ij} - h_{jk} \ge 0. \tag{10.2}$$

Let $B = L(j|j)$. By Lemma 10.1 $B^{-1} \ge 0$. We choose the following g-inverse of L: In L, replace entries in the jth row and column by zeros and replace $L(j|j)$ by B^{-1}. Let the resulting matrix be H. It is easily verified that $LHL = L$, and hence H is a g-inverse of L. Note that $h_{jj} = h_{ij} = h_{jk} = 0$, while $h_{ik} \ge 0$ since $B^{-1} \ge 0$, as remarked earlier. Thus, (10.2) is proved and the resistance distance satisfies the triangle inequality.

We make the following observation in passing. Letting H be the g-inverse as defined above, we see that

$$r(i, j) = h_{ii} + h_{jj} - h_{ij} - h_{ji} = h_{ii} = \frac{\det L(i, j|i, j)}{\det L(i|i)}. \tag{10.3}$$

The corresponding result for a tree was noted in Corollary 4.10.

10.2 Network Flows

If x is a vector of order $n \times 1$ then the norm $||x||$ is defined to be the usual Euclidean norm; $||x|| = (\sum_{i=1}^{n} x_i^2)^{\frac{1}{2}}$. We prove a preliminary result, which we will used.

Lemma 10.2 *Let A be an $n \times m$ matrix and let b be a vector of order $n \times 1$ in the column space of A. Let H be a g-inverse of A such that HA is symmetric. Then $z = Hb$ is a solution of the equation $Ax = b$ with minimum norm.*

Proof Note that $Ax = b$ is consistent, as b is in the column space of A. Let y be a solution of $Ax = b$, so that $Ay = b$. We must show $||Hb|| \leq ||y||$, or that $||HAy|| \leq ||y||$. Squaring both sides of this inequality it will be sufficient to show that $y'A'H'HAy \leq y'y$. Now

$$y'A'H'HAy = y'(HA)'HAy = y'HAHAy = y'HAy,$$

since H satisfies $AHA = A$ and $A'H' = HA$. Since HA is a symmetric, idempotent matrix, its eigenvalues are either 0 or 1, and hence $I - HA$ is positive semidefinite. It follows that $y'(I - HA)y \geq 0$ and the result is proved.   □

Let G be a connected graph with $V(G) = \{1, \ldots, n\}$ and $E(G) = \{e_1, \ldots, e_m\}$. We interpret the resistance distance between the two vertices i and j in terms of an "optimal" flow from i to j. First we give some definitions. Let the edges of G be assigned an orientation and let Q be the vertex-edge incidence matrix. A unit flow from i to j is defined as a function $f : E(G) \rightarrow \mathbb{R}$ such that

$$Q \begin{bmatrix} f(e_1) \\ f(e_2) \\ \vdots \\ f(e_m) \end{bmatrix} = e_{ij}. \tag{10.4}$$

The interpretation of (10.4) is easy: At each vertex other than i, j, the incoming flow is equal to the outgoing flow; at i the outgoing flow is 1 whereas at j, the incoming flow is also 1. The norm of a unit flow f is defined to be

$$||f|| = \left\{ \sum_{j=1}^{m} f(e_j)^2 \right\}^{\frac{1}{2}}.$$

Let L be the Laplacian matrix of G. As noted in the proof of Lemma 9.10, e_{ij} is in the column space of L, and hence in the column space of Q. Therefore, (10.4) is consistent. By Lemma 10.2, a solution of (10.4) with minimum norm is given by $f_0 = Q^- e_{ij}$, where Q^- is a minimum norm g-inverse of Q, that is, satisfies $QQ^-Q = Q$, and that Q^-Q is symmetric. Since Q^+ is a minimum

norm g-inverse of Q, $f_0 = Q^+ e_{ij}$ is a solution of (10.4) with minimum norm.
Then

$$\|f_0\|^2 = e'_{ij}(Q^+)^T Q^+ e_{ij} = e'_{ij} L^+ e_{ij}$$

since $L^+ = (QQ^T)^+ = (Q^T)^+ Q^+ = (Q^+)^T Q^+$ by well-known properties of the
Moore–Penrose inverse. Thus, we have proved that $r(i, j) = e'_{ij} L^+ e_{ij}$ is the minimum
value of $\|f\|^2$ where $\|f\|$ is a unit flow from i to j.

We illustrate the interpretation to calculate $r(u, v)$ in the following simple example.

Example 10.3 Consider the graph following:

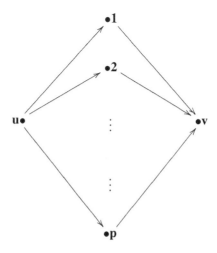

A unit flow from u to v is given by $f(\{u, k\}) = \alpha_k, f(\{k, v\}) = -\alpha_k, k = 1, 2, \ldots, p$, where $\alpha_1 + \cdots + \alpha_p = 1$. Clearly

$$\|f\|^2 = 2(\alpha_1^2 + \cdots + \alpha_p^2)$$

is minimized when $\alpha_k = \frac{1}{p}$, $k = 1, 2, \ldots, p$; in which case the value of $\|f\|^2$ is $\frac{2}{p}$.
It follows that $r(u, v) = \frac{2}{p}$.

In the next result we show that resistance distance is dominated by classical
distance.

Theorem 10.4 *Let G be a connected graph with $V(G) = \{1, \ldots, n\}$ and $E(G) = \{e_1, \ldots, e_m\}$, and let $i, j \in V(G)$. Then*

$$r(i, j) \le d(i, j). \tag{10.5}$$

Proof If $i = j$ then $r(i, j) = d(i, j) = 0$. Assume that $i \neq j$. Choose and fix an ij-path $\mathcal{P}$ of length $d(i, j)$. Orient each edge in $\mathcal{P}$ in the direction from i to j and assign an arbitrary orientation to the remaining edges of G. If $g : E(G) \rightarrow \mathbb{R}$ is defined as

$$g(e_k) = \begin{cases} 1, & \text{if } e_k \in \mathcal{P} \\ 0, & \text{otherwise} \end{cases}$$

then g is a unit flow from i to j. Since $r(i, j)$ is the minimum value of the squared norm of a unit flow from i to j, we have

$$d(i, j) = ||g||^2 \geq r(i, j),$$

and the proof is complete. $\square$

It can be shown that equality holds in (10.5) if and only if there is a unique ij-path. Before proving this statement we need a preliminary result.

Lemma 10.5 *Let G be a connected graph with $V(G) = \{1, \ldots, n\}$ and $E(G) = \{e_1, \ldots, e_m\}$. Assume that each edge of G is oriented and let Q be the vertex-edge incidence matrix. Let y be a vector of order $m \times 1$ such that $Qy = 0$. If e_k is a cut-edge of G, then $y_k = 0$.*

Proof Since e_k is a cut-edge, $G \backslash \{e_k\}$ has two components, say G_1 and G_2. We assume that e_k is oriented in the direction from $V(G_1)$ to $V(G_2)$. Let z be the incidence vector of $V(G_1)$. Thus, z is a vector of order $n \times 1$ with its components indexed by $V(G)$. The component corresponding to a vertex is 1 if it is in $V(G_1)$, and 0 otherwise. Then it can be verified that $z'Q$ is a vector of order $1 \times m$ with all the components 0 except that $z_k = 1$. It follows that $z'Qy = y_k$. Since $Qy = 0$, we conclude that $y_k = 0$. $\square$

Theorem 10.6 *Let G be a connected graph with $V(G) = \{1, \ldots, n\}$ and $E(G) = \{e_1, \ldots, e_m\}$, and let $i, j \in V(G)$. Then $r(i, j) = d(i, j)$ if there is a unique ij-path. In particular, resistance distance and classical distance coincide for a tree.*

Proof Let $g : E(G) \rightarrow \mathbb{R}$ be the unit flow from i to j as defined in the proof of Theorem 10.4. Let g also denote the $m \times 1$ vector with components $g(e_1), \ldots, g(e_m)$. Then a general unit flow from i to j is given by $g + y$, where $Qy = 0$. If there is a unique ij-path, say $\mathcal{P}$, then every edge on $\mathcal{P}$ must be a cut-edge and then, by Lemma 10.5, the components of y corresponding to the edges on $\mathcal{P}$ are zero. Thus, any unit flow from i to j coincides with g on $\mathcal{P}$. Therefore, $r(i, j)$, which equals the minimum value of the squared norm of a unit flow from i to j, must be $||g||^2 = d(i, j)$. $\square$

The converse of Theorem 10.6 is true and the proof will be left as an exercise. The fact that resistance distance and classical distance coincide for a tree is also clear from Theorem 9.12.

10.3 A Random Walk on Graphs

Let G be a connected graph with $V(G) = \{1, \ldots, n\}$ and $E(G) = \{e_1, \ldots, e_m\}$. Suppose a particle moves from vertex to vertex according to the following rule: If the particle is at the vertex k, then it moves to any of the neighbouring vertices with equal probability. Let $i, j \in V(G)$, $i \neq j$, be fixed. For $k \in V(G)$ let $P(k)$ denote the probability that a particle starting at k, and moving according to the law stated above will visit i before visiting j. Then clearly, $P(i) = 1$ and $P(j) = 0$. For $k \in V(G)$, we denote by $\mathcal{N}(k)$ the set of vertices adjacent to k. A simple argument using conditional probability shows that for $k \neq i, j$,

$$P(k) = \sum_{s \in \mathcal{N}(k)} \frac{1}{d_k} P(s), \tag{10.6}$$

where d_k denotes the degree of k. We summarize Eq. (10.6), together with $P(i) = 1$ and $P(j) = 0$, in matrix notation as follows. Let L be the Laplacian matrix of G. Let I_k denote the kth column of the identity matrix and let C be the matrix obtained from L by replacing its ith and jth rows by I_i' and I_j', respectively. Then (10.6) is equivalent to

$$CP = I_i, \tag{10.7}$$

where $P = (P(1), \ldots, P(k))'$. Since $\det C = \det L(i, j | i, j) > 0$, the system (10.7) has a unique solution. By Cramer's rule the solution is given by

$$P(k) = (-1)^{i+k} \frac{\det L(i, j | k, j)}{\det L(i, j | i, j)}, \tag{10.8}$$

for $k \neq i, j$, while $P(i) = 1$ and $P(j) = 0$.

The following identity obtained by the Laplace expansion will be useful.

$$\det L(j|j) = d_i \det L(i, j | i, j) - \sum_{k \in \mathcal{N}(i)} (-1)^{i+k} \det L(i, j | k, j). \tag{10.9}$$

Suppose the particle starts at i and moves according to the prescribed law. Let $P(i \to j)$ denote the probability that the particle visits j before returning to i. Then

$$1 - P(i \to j) = \sum_{k \in \mathcal{N}(i)} \frac{1}{d_i} P(k)$$

$$= \frac{1}{d_i} \sum_{k \in \mathcal{N}(i)} (-1)^{i+k} \frac{\det L(i, j | k, j)}{\det L(i, j | i, j)} \qquad \text{by (10.8)}$$

$$= \frac{1}{d_i} \frac{d_i \det L(i,j|i,j) - \det L(j|j)}{\det L(i,j|i,j)} \qquad \text{by (10.9)}$$

$$= 1 - \frac{1}{d_i} \frac{\det L(j|j)}{\det L(i,j|i,j)}.$$

Thus,

$$P(i \to j) = \frac{1}{d_i} \frac{\det L(j|j)}{\det L(i,j|i,j)}$$

and hence

$$r(i,j) = \frac{1}{d_i P(i \to j)}.$$

We have thus obtained an interpretation of $r(i,j)$ in terms of the random walk on G. The interpretation is justified intuitively. If vertices i and j are far apart then a particle starting at i is more likely to return to i before it visits j.

There is a close connection between the interpretation of $r(i,j)$ in terms of the random walk and the one based on electrical networks, which we discuss in the next section.

10.4 Effective Resistance in Electrical Networks

Let G be a connected graph with $V(G) = \{1, \ldots, n\}$, and let $i,j \in V(G), i \neq j$. We think of G as an electrical network in which a unit resistance is placed along each edge. Current is allowed to enter the network only at vertex i and leave it only at j. Let $v(k)$ denote the voltage at the vertex k. We set $v(i) = 1$ and $v(j) = 0$. By Ohm's law, the current flowing from x to y, where $\{xy\} \in E(G)$, is given by $v(x) - v(y)$. According to Kirchhoff's law, at any point $k \in V(G), k \neq i, j$,

$$\sum_{y \in \mathcal{N}(k)} (v(k) - v(y)) = 0.$$

If we set $v = (v(1), \ldots, v(n))'$, then v satisfies the equation

$$Cv = e_i, \qquad (10.10)$$

where C is precisely the matrix defined in the previous section. As in the previous section, the solution of (10.10) is given by

$$v(k) = (-1)^{i+k} \frac{\det L(i,j|k,j)}{\det L(j|j)} \qquad (10.11)$$

for $k \neq j$ and $v(j) = 0$. The current flowing into the network at vertex i is given by the sum of the currents from y to i for each $y \in \mathcal{N}(i)$ and this equals

$$\sum_{y\in\mathcal{N}(i)} (v(y) - v(i)) = \sum_{y\in\mathcal{N}(i)} (v(y) - 1) = \sum_{y\in\mathcal{N}(i)} v(y) - d_i.$$

Carrying out this calculation using (10.10), (10.11) as in the previous section we find that the current flowing into the network is

$$\frac{\det L(j|j)}{\det L(i,j|i,j)},$$

which is precisely the reciprocal of $r(i,j)$, in view of (10.3). The reciprocal of the current is called the "effective resistance" between i,j and this justifies the term "resistance distance".

The standard techniques of finding resistance in an electrical network, such as series-parallel reduction, may be employed to find resistance distance. We leave it as an exercise to verify that the resistance distance between i and j in the following graph is $\frac{4}{9}$:

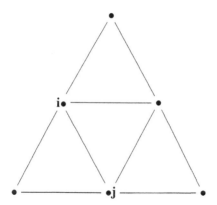

10.5 Resistance Matrix

Let G be a connected graph with $V(G) = \{1, \ldots, n\}$. The resistance matrix R of G is an $n \times n$ matrix defined as follows. The rows and the columns of R are indexed by $V(G)$. For $i, j \in \{1, \ldots, n\}$, the (i, j)-entry of R is defined as $r_{ij} = r(i,j)$, the resistance distance between i and j. When G is a tree R reduces to the distance matrix D of the tree. We show that certain formulas involving the distance matrix of a tree extend naturally to the case of the resistance matrix. These include a formula for the inverse of the resistance matrix.

We introduce some notation. Let L be the Laplacian of G. By Lemma 4.5 the eigenvalues of $L + \frac{1}{n}J$ are positive and hence the matrix is nonsingular. We set

$$X = \left(L + \frac{1}{n}J\right)^{-1}.$$

It is easily verified, using $X\left(L + \frac{1}{n}J\right) = \left(L + \frac{1}{n}J\right)X = I$, that

$$L^+ = X - \frac{1}{n}J.$$

Let $\tilde{X}$ be the diagonal matrix $\text{diag}(x_{11}, \ldots, x_{nn})$. With this notation we have the following:

Lemma 10.7 $R = \tilde{X}J + J\tilde{X} - 2X$.

Proof The (i, j)-element of $\tilde{X}J + J\tilde{X} - 2X$ equals

$$x_{ii} + x_{jj} - 2x_{ij} = \ell_{ii}^+ + \ell_{jj}^+ - 2\ell_{ij}^+,$$

since $L^+ = X - \frac{1}{n}J$. The result follows by the definition of resistance distance.   □

For $i = 1, \ldots, n$, let

$$\tau_i = 2 - \sum_{j \sim i} r(i, j).$$

Let τ be the $n \times 1$ vector with components $\tau_1, \ldots, \tau_n$.

Lemma 10.8 $\tau = L\tilde{X}\mathbf{1} + \frac{2}{n}\mathbf{1}$.

Proof Let d_i be the degree of vertex i, $i = 1, \ldots, n$. Since $\left(L + \frac{1}{n}J\right)X = I$, we have

$$d_i x_{ii} - \sum_{j \sim i} x_{ij} + \frac{1}{n}\sum_{j=1}^{n} x_{ij} = 1, \quad i = 1, \ldots, n. \tag{10.12}$$

The row sums of $L + \frac{1}{n}J$ are all 1 and hence the row sums of X are 1 as well. It follows from (10.12) that

$$d_i x_{ii} - \sum_{j \sim i} x_{ij} = 1 - \frac{1}{n}, \quad i = 1, \ldots, n. \tag{10.13}$$

For $i = 1, \ldots, n$,

$$\tau_i = 2 - \sum_{j \sim i} r(i, j)$$

$$= 2 - \sum_{j \sim i} (x_{ii} + x_{jj} - 2x_{ij}) \qquad\qquad \text{by Lemma 10.7}$$

$$= 2 - \sum_{j \sim i} x_{ii} - \sum_{j \sim i} x_{jj} + 2\sum_{j \sim i} x_{ij}$$

$$= 2 - d_i x_{ii} - \sum_{j \sim i} x_{jj} + 2 \sum_{j \sim i} x_{ij}$$

$$= 2 - d_i x_{ii} - \sum_{j \sim i} x_{jj} + \left(2d_i x_{ii} - 2 + \frac{2}{n} \right) \qquad \text{by (10.13)}$$

$$= d_i x_{ii} - \sum_{j \sim i} x_{jj} + \frac{2}{n},$$

which is clearly the ith entry of $L\tilde{X}\mathbf{1} + \frac{2}{n}\mathbf{1}$. Hence, the proof is complete. □

Lemma 10.9 $\sum_{i=1}^{n} \sum_{j \sim i} r(i,j) = 2(n-1)$.

Proof Recall that $LL^+ = I - \frac{1}{n}J$. Also, since the row sums of L are zero, $LX = LL^+$. By Lemma 10.7,

$$\begin{aligned}
LR &= L(\tilde{X}J + J\tilde{X} - 2X) \\
&= L\tilde{X}J - 2LX \\
&= L\tilde{X}J - 2LL^+ \\
&= L\tilde{X}J - 2\left(I - \frac{1}{n}J \right).
\end{aligned} \qquad (10.14)$$

It is easily verified that

$$\sum_{i=1}^{n} \sum_{j \sim i} r(i,j) = -\text{trace } LR. \qquad (10.15)$$

It follows from (10.14) and (10.15) that

$$\begin{aligned}
\sum_{i=1}^{n} \sum_{j \sim i} r(i,j) &= -\text{trace } LR \\
&= -\text{trace } L\tilde{X}J + 2(n-1) \\
&= -\text{trace } L\tilde{X}\mathbf{11}' + 2(n-1) \\
&= -\mathbf{1}'L\tilde{X}\mathbf{1} + 2(n-1) \\
&= 2(n-1),
\end{aligned}$$

and the proof is complete. □

The next result is an extension of Lemma 9.6.

Corollary 10.10 $\mathbf{1}'\tau = 2$.

Proof By Lemma 10.9,

$$\mathbf{1}'\tau = 2n - \sum_{i=1}^{n}\sum_{j\sim i} r(i,j) = 2n - 2(n-1) = 2,$$

and the result is proved.   □

Let $\tilde{x}$ denote the $n \times 1$ vector whose components are the diagonal elements of $\tilde{X}$.

Lemma 10.11 $\tau'R\tau = 2\tilde{x}'L\tilde{x} + \frac{8}{n}$ *trace* (L^+).

Proof By Lemma 10.8,

$$\tau'R\tau = \left(\mathbf{1}'\tilde{X}L + \frac{2}{n}\mathbf{1}'\right) R \left(L\tilde{X}\mathbf{1} + \frac{2}{n}\mathbf{1}\right)$$

$$= \mathbf{1}'\tilde{X}LRL\tilde{X}\mathbf{1} + \frac{4}{n}\mathbf{1}'\tilde{X}LR\mathbf{1} + \frac{4}{n^2}\mathbf{1}'R\mathbf{1}. \qquad (10.16)$$

By Lemma 10.7,

$$LRL = L(\tilde{X}J + J\tilde{X} - 2X)L$$
$$= -2LXL = -2LL^+L = -2L. \qquad (10.17)$$

It follows from (10.17) that

$$\mathbf{1}'\tilde{X}LRL\tilde{X}\mathbf{1} = -2\tilde{x}'L\tilde{x}. \qquad (10.18)$$

Again, using (10.14) we get
$$\mathbf{1}'\tilde{X}LR\mathbf{1} = n\tilde{x}'L\tilde{x}. \qquad (10.19)$$

Finally, using Lemma 10.7 and the fact that X has row sums 1,

$$\mathbf{1}'R\mathbf{1} = 2n\,\text{trace}\ (X) - 2n = 2n\,\text{trace}\ (L^+). \qquad (10.20)$$

The result follows from (10.17), (10.18), (10.19) and (10.20).   □

The next result generalizes the formula for the inverse of the distance matrix of a tree, obtained in Theorem 9.9.

Theorem 10.12 $R^{-1} = -\frac{1}{2}L + \frac{1}{\tau'R\tau}\tau\tau'$.

Proof It follows from Lemma 10.7 and (10.14) that

$$LR + 2I = L\tilde{X}J + \frac{2}{n}J = \tau\mathbf{1}'. \qquad (10.21)$$

Using (10.21) and Corollary 10.10 we have

$$(LR + 2I)\tau = \tau \mathbf{1}'\tau = 2\tau,$$

and hence $LR\tau = 0$. From Lemma 10.11 we conclude that $R\tau$ is a nonzero vector and then, since $LR\tau = 0$, there must be a nonzero scalar α such that $R\tau = \alpha \mathbf{1}$. Then, by Corollary 10.10, $\tau'R\tau = \alpha\tau'\mathbf{1} = 2\alpha$ and hence $\alpha = \frac{\tau'R\tau}{2}$. Therefore,

$$R\tau = \frac{\tau'R\tau}{2}\mathbf{1}. \tag{10.22}$$

It follows from (10.21) and (10.22) that

$$\left(-\frac{1}{2}L + \frac{1}{\tau'R\tau}\tau\tau'\right)R = -\frac{1}{2}LR + \frac{1}{\tau'R\tau}\tau\tau'R$$

$$= I - \frac{1}{2}\tau\mathbf{1}' + \frac{1}{\tau'R\tau}\left(\frac{\tau'R\tau}{2}\right)\tau\mathbf{1}'$$

$$= I,$$

and the proof is complete.   □

Exercises

1. Let G be a connected graph with $V(G) = \{1, \ldots, n\}$, let L be the Laplacian of G and let $\lambda_1 \geq \cdots \geq \lambda_{n-1} > \lambda_n = 0$ be the eigenvalues of L. Show that

$$\sum_{i=1}^{n}\sum_{j=1}^{n} r(i,j) = 2n\sum_{i=1}^{n-1}\frac{1}{\lambda_i}.$$

2. Let C_n be the cycle on the n vertices $\{1, \ldots, n\}$. Show that for $i = 1, \ldots, n$,

$$\sum_{j\sim i} r(i,j) = 2 - \frac{2}{n}.$$

3. Let G be a connected graph and let i, j be distinct vertices of G. If $r(i,j) = d(i,j)$, show that there is a unique ij-path.

4. Show that the resistance matrix of a connected graph on $n \geq 2$ vertices has exactly one positive eigenvalue.

5. Let G be a connected graph with $V(G) = \{1, \ldots, n\}$. Let $i, j \in V(G)$ and suppose that an ij-path contains k, which is a cut-vertex. Show that $r(i,j) = r(i,k) + r(k,j)$.

6. Let G be a connected graph with $V(G) = \{1, \ldots, n\}$. Let $i, j \in V(G)$ be adjacent vertices, joined by the edge e_k. Let $\kappa(G)$ be the number of spanning trees in G and let $\kappa'(G)$ be the number of spanning trees in G containing e_k. Show that $r(i, j) = \frac{\kappa'(G)}{\kappa(G)}$.

7. Let G be a planar graph and let G^* be the dual graph of G. Let e_k be an edge of G with endpoints u, v, and let e'_k be the corresponding edge in G^* with endpoints u', v'. If $r(u, v)$ is the resistance distance between u, v in G, and $r'(u', v')$ is the resistance distance between u', v' in G^*, show that

$$r(u, v) + r'(u', v') = 1.$$

8. Let G be a connected graph with $V(G) = \{1, \ldots, n\}$. Show that

$$\sum_{i=1}^{n} \sum_{j \sim i} r(i, j) = 2(n - 1).$$

9. Let G be a connected graph with n vertices, R be the resistance matrix of G, τ be as defined after Lemma 10.7 and $\kappa(G)$ be the number of spanning trees in G. Show that

$$\det R = (-1)^{n-1} 2^{n-3} \frac{\tau' R \tau}{\kappa(G)}.$$

10. Let T be a tree with $V(T) = \{1, \ldots, n\}$, D be the distance matrix of T, and $\tau_i = 2 - d_i$, where d_i is the degree of vertex i, $i = 1, \ldots, n$. Let τ be the $n \times 1$ vector with components $\tau_1, \ldots, \tau_n$. Show that $\tau' D \tau = 2(n-1)$. Hence, conclude that Theorem 9.9 is a special case of Theorem 10.12.

The term "resistance distance" was introduced by Klein and Randić [KR93]. The treatment in this chapter is based on [B99, B04], where further references can be found. Bollobás [B98] and Doyle and Snell [DS84] are classical references for a graph theoretic treatment of resistance.

References and Further Reading

[B99]   Bapat, R.B.: Resistance distance in graphs. Math. Student **68**, 87–98 (1999)
[B04]   Bapat, R.B.: Resistance matrix of a weighted graph. MATCH Commun. Math. Comput. Chem. **50**, 73–82 (2004)
[B98]   Bollobás, B.: Modern Graph Theory. Springer-Verlag, New York (1998)
[DS84]   Doyle, P.G., Snell, J.L.: Random Walks and Electrical Networks. Math. Assoc. Am, Washington (1984)
[KR93]   Klein, D.J., Randić, M.: Resistance distance. J. Math. Chem. **12**, 81–95 (1993)

Chapter 11
Laplacian Eigenvalues of Threshold Graphs

Threshold graphs have an interesting structure and they arise in many areas. We will be particularly interested in the Laplacian eigenvalues of threshold graphs. We first review some basic aspects of the theory of majorization.

11.1 Majorization

If $x \in \mathbb{R}^n$, let $x_{[1]} \geq \cdots \geq x_{[n]}$ be a rearrangement of the coordinates of x in nonincreasing order. If $x, y \in \mathbb{R}^n$ then x is said to be *majorized* by y, denoted $x \prec y$, if the following conditions hold:

$$\sum_{i=1}^{k} x_{[i]} \leq \sum_{i=1}^{k} y_{[i]}, \quad i = 1, \ldots, n-1, \tag{11.1}$$

and

$$\sum_{i=1}^{n} x_i = \sum_{i=1}^{n} y_i. \tag{11.2}$$

If $x \prec y$, then, intuitively, coordinates of x are less "spread out" than coordinates of y. As an example, $[2, 3, 2, 3]'$ is majorized by $[5, 1, 1, 3]'$. If $x \in \mathbb{R}^n$ and if $\bar{x}$ is the arithmetic mean of $x_1, \ldots, x_n$, then it can be verified that $[\bar{x}, \ldots, \bar{x}]'$ is majorized by x. If x and y are $1 \times n$ vectors then we say that $x \prec y$ if $x' \prec y'$. For $x, y \in \mathbb{R}^n$, if $x \prec y$ we often say that $x_1, \ldots, x_n$ are majorized by $y_1, \ldots, y_n$.

Let A be an $n \times n$ matrix. Recall that A is said to be doubly stochastic if $a_{ij} \geq 0$ for all i, j, and the row and the column sums of A are all equal to 1. The next result is the Hardy–Littlewood–Polya theorem. We prove only the sufficiency part.

Theorem 11.1 *Let $x, y \in \mathbb{R}^n$. Then $x \prec y$ if and only if there exists an $n \times n$ doubly stochastic matrix A such that $x = Ay$.*

© Springer-Verlag London 2014
R.B. Bapat, *Graphs and Matrices*, Universitext,
DOI 10.1007/978-1-4471-6569-9_11

Proof (Sufficiency) Let A be an $n \times n$ doubly stochastic matrix such that $x = Ay$. Clearly,

$$\sum_{i=1}^{n} x_i = \sum_{i=1}^{n} \left(\sum_{j=1}^{n} a_{ij} y_j \right) = \sum_{j=1}^{n} y_j \left(\sum_{i=1}^{n} a_{ij} \right) = \sum_{j=1}^{n} y_j. \qquad (11.3)$$

Let k be fixed, $1 \leq k \leq n - 1$. We assume, without loss of generality, that $x_1 \geq \cdots \geq x_n$ and $y_1 \geq \cdots \geq y_n$, since this ordering only amounts to permuting rows and columns of A, which again results in a doubly stochastic matrix. Let

$$t_j = \sum_{i=1}^{k} a_{ij}, \quad j = 1, \ldots, n.$$

Note that

$$\sum_{j=1}^{n} t_j = \sum_{j=1}^{n} \sum_{i=1}^{k} a_{ij} = \sum_{i=1}^{k} \sum_{j=1}^{n} a_{ij} = k.$$

We have

$$\sum_{i=1}^{k} (x_i - y_i) = \sum_{i=1}^{k} \left(\sum_{j=1}^{n} a_{ij} y_j \right) - \sum_{i=1}^{k} y_i$$

$$= \sum_{j=1}^{n} y_j \left(\sum_{i=1}^{k} a_{ij} \right) - \sum_{i=1}^{k} y_i$$

$$= \sum_{j=1}^{n} y_j t_j - \sum_{i=1}^{k} y_i + y_k \left(k - \sum_{i=1}^{n} t_i \right)$$

$$= \sum_{j=1}^{k} (y_j - y_k)(t_j - 1) + \sum_{j=k+1}^{n} t_j (y_j - y_k)$$

$$\leq 0.$$

Hence,

$$\sum_{i=1}^{k} x_i \leq \sum_{i=1}^{k} y_i, \quad k = 1, \ldots, n - 1. \qquad (11.4)$$

It follows from (11.3) and (11.4) that $x \prec y$ and the proof is complete. $\qquad \square$

An important consequence of Theorem 11.1 is stated in the next result.

Theorem 11.2 *Let A be a symmetric $n \times n$ matrix and let $\lambda_1, \ldots, \lambda_n$ be the eigenvalues of A. Then*

$$(a_{11}, \ldots, a_{nn}) \prec (\lambda_1, \ldots, \lambda_n).$$

Proof By the spectral theorem there exists an orthogonal matrix P such that

$$A = P \begin{bmatrix} \lambda_1 & \cdots & 0 \\ \vdots & \ddots & \vdots \\ 0 & \cdots & \lambda_n \end{bmatrix} P'.$$

Hence,

$$a_{ii} = \sum_{j=1}^{n} p_{ij}^2 \lambda_j, \quad i = 1, \ldots, n. \tag{11.5}$$

Since P is orthogonal, it follows that the $n \times n$ matrix with (i, j)-element p_{ij}^2 is doubly stochastic. The result follows from (11.5) and Theorem 11.2. $\square$

Corollary 11.3 *Let G be a graph with $V(G) = \{1, \ldots, n\}$. Let L be the Laplacian of G and $\lambda_1, \ldots, \lambda_n$ be the eigenvalues of L. If $d_1, \ldots, d_n$ are the vertex degrees, then*

$$(d_1, \ldots, d_n) \prec (\lambda_1, \ldots, \lambda_n).$$

We now consider the majorization relation between vectors of integer coordinates. Let $b_1, \ldots, b_n$ be integers and suppose $b_i > b_j$ for some i, j. Define

$$b_i' = b_i - 1, \quad b_j' = b_j + 1$$

and

$$b_k' = b_k, \quad k \neq i, j.$$

We say that $b_1', \ldots, b_n'$ are obtained from $b_1, \ldots, b_n$ by a transfer, or, more specifically, a transfer from i to j. We say that the vector b is obtained from the vector a by a transfer if the coordinates of b are obtained from those of a by a transfer.

Theorem 11.4 *Let a, b be $n \times 1$ vectors of integers. Then $a \prec b$ if and only if a is obtained from b by a finite number of transfers.*

Proof First, suppose that a is obtained from b by a single transfer. Then clearly the sum of the largest k elements from $b_1, \ldots, b_n$ cannot be less than the sum of the largest k elements from $a_1, \ldots, a_n$, $k = 1, \ldots, n - 1$. It is also obvious that $\sum_{i=1}^n a_i = \sum_{i=1}^n b_i$ and hence $a \prec b$. By a repeated application of this observation we conclude that $a \prec b$ when a is obtained from b by a finite sequence of transfers.

To prove the converse, assume $a \prec b$, $a \neq b$, and, without loss of generality, let $a_1 \geq \cdots \geq a_n$ and $b_1 \geq \cdots \geq b_n$. Let ℓ be the largest integer for which

$$\sum_{i=1}^{\ell} a_i < \sum_{i=1}^{\ell} b_i.$$

Then $a_{\ell+1} > b_{\ell+1}$, and there is a largest integer $p < \ell$ for which $a_p < b_p$. Thus,

$$b_p > a_p > a_{\ell+1} > b_{\ell+1}.$$

Let b' be obtained from b by a transfer from p to $\ell+1$. Then $a \prec b' \prec b$. Continuing this process we see that a is obtained from b by a finite number of transfers. □

Let $a_1, \ldots, a_n$ be nonnegative integers. Define

$$a_j^* = |\{a_i : a_i \geq j\}|, \quad j = 1, 2, \ldots$$

Thus, a_j^* is the number of a_i that are greater than or equal to j. We say that the sequence $a_1^*, a_2^*, \ldots$ is conjugate to (or the conjugate sequence of) $a_1, a_2, \ldots, a_n$. Often we may ignore some, or all, of the zeros in the two sequences. As an example, $7, 5, 4, 3, 2$ is the conjugate sequence of $5, 5, 4, 3, 2, 1, 1$.

It is instructive to consider another interpretation of conjugate sequence. Let $a_1 \geq \cdots \geq a_n$ be nonnegative integers. The Ferrers diagram corresponding to $a_1, \ldots, a_n$ consists of n left-justified rows of boxes, where the ith row consists of a_i boxes, $i = 1, \ldots, n$. If $a_i = 0$, the ith row is absent. For example, the Ferrers diagram corresponding to $5, 3, 3, 3, 2, 1$ is

Let $a_1 \geq \cdots \geq a_n$ be nonnegative integers and consider the corresponding Ferrers diagram. Then note that a_i^* is the number of boxes in the ith column of the Ferrers diagram, $i = 1, \ldots, n$. As an immediate consequence of this observation we see that if $a_1^*, \ldots, a_m^*$ is the conjugate sequence of $a_1, \ldots, a_n$, then

$$\sum_{i=1}^{n} a_i = \sum_{i=1}^{n} a_i^*.$$

We now state the Gale–Ryser Theorem. We prove only the necessity.

Theorem 11.5 *Let* $r_1 \geq \cdots \geq r_m$ *and* $c_1 \geq \cdots \geq c_n$ *be nonnegative integers such that* $r_i \leq n$, $i = 1, \ldots, m$, *and* $\sum_{i=1}^{m} r_i = \sum_{i=1}^{n} c_i$. *Then there exists an* $m \times n$ $(0 - 1)$-*matrix* A *with row sums* $r_1, \ldots, r_m$ *and column sums* $c_1, \ldots, c_n$ *if and only if* $c_1, \ldots, c_n$ *is majorized by* $r_1^*, \ldots, r_n^*$.

Proof (necessity) Let A be an $m \times n(0 - 1)$-matrix with row sums $r_1, \ldots, r_m$ and column sums $c_1, \ldots, c_n$. We assume, without loss of generality, that $c_1 \geq \cdots \geq c_n$.

Suppose there exist i, j such that $a_{ij} = 0$ and $a_{ij+1} = 1$. Let B be the matrix defined as

$$b_{ij} = 1, \quad b_{ij+1} = 0,$$

and $b_{k\ell} = a_{k\ell}$, otherwise. If $c'_1 \ldots, c'_n$ are the column sums of B then $c'_j = c_j + 1, c'_{j+1} = c_{j+1} - 1$ and $c'_k = c_k$, $k \neq j$, $j + 1$. Thus, $c_1, \ldots, c_n$ can be obtained from $c'_1 \ldots, c'_n$ by a transfer from $j + 1$ to j. It follows by Theorem 11.4 that $c_1, \ldots, c_n$ is majorized by $c'_1, \ldots, c'_n$. Continuing this process we obtain the $m \times n$ matrix C whose row sums are $r_1, \ldots, r_m$, in which row i consists of r_i 1s followed by zeros, and whose column sums majorize $c_1, \ldots, c_n$. As seen in the context of Ferrers diagram, the column sums of C are $r_1^*, \ldots, r_n^*$ and the result follows. $\square$

Consider the matrix

$$A = \begin{bmatrix} 0 & 1 & 1 & 0 & 1 & 0 & 1 & 1 \\ 1 & 0 & 1 & 1 & 1 & 1 & 1 & 0 \\ 1 & 0 & 0 & 1 & 1 & 1 & 1 & 0 \\ 1 & 1 & 0 & 0 & 0 & 1 & 0 & 1 \\ 1 & 0 & 1 & 0 & 0 & 0 & 0 & 1 \end{bmatrix}.$$

The row sums of A are $5, 6, 5, 4, 3$ and the column sums are $4, 2, 3, 2, 3, 3, 3, 3$. The conjugate sequence of the row sum sequence is $5, 5, 5, 4, 3, 1, 0, 0$, which clearly majorizes the sequence of column sums. It may be remarked that in this example the matrix C constructed in the proof of Theorem 11.5 is

$$C = \begin{bmatrix} 1 & 1 & 1 & 1 & 1 & 0 & 0 & 0 \\ 1 & 1 & 1 & 1 & 1 & 1 & 0 & 0 \\ 1 & 1 & 1 & 1 & 1 & 0 & 0 & 0 \\ 1 & 1 & 1 & 1 & 0 & 0 & 0 & 0 \\ 1 & 1 & 1 & 0 & 0 & 0 & 0 & 0 \end{bmatrix},$$

and it has the column sums $5, 5, 5, 4, 3, 1, 0, 0$, which is the conjugate sequence of the sequence of column sums.

Corollary 11.6 *Let G be a graph with $V(G) = \{1, \ldots, n\}$. Let $d_1, \ldots, d_n$ be the degree sequence of G. Then, $d_1, \ldots, d_n$ is majorized by $d_1^*, \ldots, d_n^*$.*

Proof Let A be the adjacency matrix of G. Then, A is a $(0 - 1)$-matrix and the row sums as well as the column sums of A are $d_1, \ldots, d_n$. The result follows from Theorem 11.5. $\square$

11.2 Threshold Graphs

Threshold graphs are best defined using a recursive procedure. A vertex is called *dominating* if it is adjacent to every other vertex. A graph G with $V(G) = \{1, \ldots, n\}$ is called a *threshold graph* if it is obtained by the following procedure. Start with K_1, a single vertex, and use any of the following steps, in any order, an arbitrary number of times:

(i) Add an isolated vertex.
(ii) Add a dominating vertex, that is, add a new vertex and make it adjacent to each existing vertex.

For example, the star $K_{1,n}$ is a threshold graph. The following graphs are also threshold:

Given a graph G, we have the following recursive procedure to check whether G is a threshold graph:

(i) If G is connected then in order for it to be threshold, it necessarily has a dominating vertex. After deleting that vertex the connected components of the resulting graph must consist of a single connected component, say H, together with possibly some isolated vertices. Furthermore, G is threshold if and only if H is so. We check whether H is threshold.
(ii) If G is disconnected then in order for it to be threshold it necessarily has a single connected component, say K, together with possibly some isolated vertices. Furthermore, G is threshold if and only if K is so. We check whether K is threshold.

We now prove a preliminary result.

Lemma 11.7 *Let G be a graph with $V(G) = \{1, \ldots, n\}$. Let $d_1 \geq \cdots \geq d_n$ be the degree sequence of G and suppose $d_1 = n - 1$. Let $H = G \setminus \{1\}$ and $L(G)$ and $L(H)$ be the Laplacians of G and H, respectively. Then n is an eigenvalue of $L(G)$. Furthermore, if $\lambda_2, \ldots, \lambda_{n-1}, n$ and 0 are the eigenvalues of $L(G)$, then the eigenvalues of $L(H)$ are $\lambda_2 - 1, \ldots, \lambda_{n-1} - 1$ and 0.*

Proof Note that

$$
L(G) + J_n =
\begin{bmatrix}
n & 0 & \cdots & & 0 \\
0 & & & & \\
\vdots & & L(H) + J_{n-1} + I_{n-1} & \\
0 & & & &
\end{bmatrix}.
\tag{11.6}
$$

By Lemma 4.5 the eigenvalues of $L(G)+J_n$ are $\lambda_2, \ldots, \lambda_{n-1}$, and n with multiplicity 2. By (11.6), the eigenvalues of $L(H)+J_{n-1}+I_{n-1}$ are $\lambda_2, \ldots, \lambda_{n-1}$ and n. Therefore, the eigenvalues of $L(H) + J_{n-1}$ are $\lambda_2 - 1, \ldots, \lambda_{n-1} - 1$ and $n - 1$. It follows from Lemma 4.5 that the eigenvalues of $L(H)$ are $\lambda_2 - 1, \ldots, \lambda_{n-1} - 1$ and 0, and the proof is complete. $\square$

The Laplacian eigenvalues of a threshold graph enjoy an interesting property, which is proved next.

Theorem 11.8 *Let G be a threshold graph with $V(G) = \{1, \ldots, n\}$. Let $L(G)$ be the Laplacian and $d_1, \ldots, d_n$ the degree sequence of G. Then $d_1^*, \ldots, d_n^*$ are the eigenvalues of $L(G)$.*

Proof The result will be proved by induction on n. For $n = 1$ the result is trivial. Assume the result to be true for threshold graphs of at most $n - 1$ vertices and consider G, which is a threshold graph with n vertices. Clearly, it will be sufficient to prove the result for a connected threshold graph, since an isolated vertex contrbutes a 0 to the degree seqence, as well as to the Laplacian eigenvalues. So we assume that G is connected. It follows from the definition of a threshold graph that G has a dominating vertex, which we assume to be 1. Let $H = G \setminus \{1\}$ and let $L(H)$ be the Laplacian of H. Let $\lambda_1, \ldots, \lambda_{n-1}$ and $\lambda_n = 0$ be the eigenvalues of $L(G)$. By Lemma 11.7 n is an eigenvalue of $L(G)$, and we assume that $\lambda_1 = n$. By Lemma 11.7 the eigenvalues of $L(H)$ are $\lambda_2 - 1, \ldots, \lambda_{n-1} - 1$, and 0. Observe that if we add k isolated vertices to a graph then both the degree sequence as well as the eigenvalues get augmented by k zeros. Thus, if the Laplacian eigenvalues of a graph are given by the conjugate of its degree sequence then this property continues to hold when some isolated vertices are added. Since the connected components of H consist of a threshold graph, and possibly some isolated vertices, by the induction assumption, $\lambda_2 - 1, \ldots, \lambda_{n-1} - 1, 0$ is the conjugate sequence of the degree sequence of H, which is $d_2 - 1, \ldots, d_n - 1$. Since $\lambda_1 = n$ and $d_1 = n - 1$, it follows that $\lambda_1, \ldots, \lambda_{n-1}, 0$ is the conjugate sequence of $d_1, \ldots, d_n$ and the proof is complete.   □

The converse of Theorem 11.8 is also true and is stated next. The proof, which is similar to that of Theorem 11.8, will be left as an exercise.

Theorem 11.9 *Let G be a connected graph with $V(G) = \{1, \ldots, n\}$. Let $L(G)$ be the Laplacian and $d_1, \ldots, d_n$ the degree sequence of G. If $d_1^*, \ldots, d_n^*$ are the eigenvalues of $L(G)$ then G is a threshold graph.*

11.3 Spectral Integral Variation

A graph is called *Laplacian integral* if the eigenvalues of its Laplacian are all integers. Threshold graphs are Laplacian integral. Besides threshold graphs there are other Laplacian integral graphs as well. As an example, we describe another class of Laplacian integral graphs, which includes threshold graphs. A graph is called a cograph if it is constructed using the following rules:

  (i) K_1 is a cograph.
 (ii) The complement of a cograph is a cograph.
(iii) The union of two vertex-disjoint cographs is a cograph.

Note that the definition gives a recursive procedure to construct a cograph. We may take the union of two vertex disjoint cographs. Then its complement is again a cograph.

It is easy to see that a cograph is Laplacian integral, the proof of which will be left as an exercise. It is also clear that threshold graphs are cographs. However, the converse is not true. The cycle C_4 is a cograph, but not a threshold graph.

We now obtain some results concerning the effect of a rank 1 perturbation on the eigenvalues of a symmetric matrix. The results will be applied to examine the change in the Laplacian eigenvalues of a graph, when a single edge is added to the graph. We first prove some preliminary results.

Lemma 11.10 *Let A be a symmetric $n \times n$ matrix partitioned as*

$$A = \begin{bmatrix} a_{11} & x' \\ x & A(1|1) \end{bmatrix}.$$

If the eigenvalues of $A(1|1)$ consist of $n - 1$ eigenvalues of A then $x = 0$.

Proof Let $\mu_1, \ldots, \mu_n$ be the eigenvalues of A and suppose the eigenvalues of $A(1|1)$ are

$$\mu_1, \ldots, \mu_{k-1}, \mu_{k+1}, \ldots, \mu_n,$$

where $1 \le k \le n$. Then

$$\text{trace } A - \text{trace} A(1|1) = \mu_k, \quad \text{trace } A^2 - \text{trace } A(1|1)^2 = \mu_k^2. \tag{11.7}$$

Using the partition of A,

$$\text{trace } A - \text{trace} A(1|1) = a_{11}, \quad \text{trace } A^2 = a_{11}^2 + 2x'x + \text{trace } A(1|1)^2. \tag{11.8}$$

It follows from (11.7) and (11.8) that $a_{11} = \mu_k$, and hence $x'x = 0$. Therefore, $x = 0$ and the proof is complete. $\qquad\square$

Lemma 11.11 *Let A be a symmetric $n \times n$ matrix partitioned as*

$$A = \begin{bmatrix} a_{11} & x' \\ x & A(1|1) \end{bmatrix}$$

and let

$$B = \begin{bmatrix} a_{11} + \beta & x' \\ x & A(1|1) \end{bmatrix},$$

where $\beta \ne 0$. Suppose the eigenvalues of A are $\mu_1, \ldots, \mu_n$ and the eigenvalues of B are $\mu_1, \ldots, \mu_{k-1}, \mu_k + \beta, \mu_{k+1}, \ldots, \mu_n$ for some k, $1 \le k \le n$. Then $x = 0$.

Proof The characteristic polynomials of A and B are

$$\det(\lambda I - A) = (\lambda - \mu_1) \cdots (\lambda - \mu_n)$$

and

$$\det(\lambda I - B) = (\lambda - \mu_1) \cdots (\lambda - \mu_k - \beta) \cdots (\lambda - \mu_n),$$

respectively. Hence,

$$\det(\lambda I - B) = \det(\lambda I - A) - \beta(\lambda - \mu_1) \cdots (\lambda - \mu_{k-1})(\lambda - \mu_{k+1}) \cdots (\lambda - \mu_n). \quad (11.9)$$

From the partition of B, we have

$$\det(\lambda I - B) = \det(\lambda I - A) - \beta \det(\lambda I - A(1|1)). \quad (11.10)$$

Since $\beta \neq 0$, it follows from (11.9) and (11.10) that

$$\det(\lambda I - A(1|1)) = (\lambda - \mu_1) \cdots (\lambda - \mu_{k-1})(\lambda - \mu_{k+1}) \cdots (\lambda - \mu_n).$$

This implies that the eigenvalues of $A(1|1)$ consist of $n - 1$ eigenvalues of A, and it follows from Lemma 11.10 that $x = 0$. □

Theorem 11.12 *Let A be a symmetric $n \times n$ matrix with eigenvalues $\mu_1, \ldots, \mu_n$. Let B be a symmetric $n \times n$ matrix of rank 1, and let β be the nonzero eigenvalue of B. Then the eigenvalues of $A + B$ are $\mu_1, \ldots, \mu_{k-1}, \mu_k + \beta, \mu_{k+1}, \ldots, \mu_n$ for some $k, 1 \leq k \leq n$ if and only if $AB = BA$.*

Proof If $AB = BA$ then A and B can be simultaneously diagonalized. Thus, there exists an orthogonal P such that PAP' and PBP' are both diagonal with the eigenvalues of A along the diagonal and the eigenvalues of B along the diagonal, respectively. Then $P(A + B)P'$ is diagonal with the diagonal entries equal to $\mu_1, \ldots, \mu_{k-1}, \mu_k + \beta, \mu_{k+1}, \ldots, \mu_n$ for some $k, 1 \leq k \leq n$. This proves the "if" part.

To prove the converse we may assume, using the spectral theorem, that $B = \text{diag}(\beta, 0, \ldots, 0)$. Let A be partitioned as

$$A = \begin{bmatrix} a_{11} & x' \\ x & A(1|1) \end{bmatrix}.$$

Then

$$A + B = \begin{bmatrix} a_{11} + \beta & x' \\ x & A(1|1) \end{bmatrix}$$

and has eigenvalues $\mu_1, \ldots, \mu_{k-1}, \mu_k + \beta, \mu_{k+1}, \ldots, \mu_n$ for some $k, 1 \leq k \leq n$. It follows from Lemma 11.11 that $x = 0$. Then

$$A = \begin{bmatrix} a_{11} & 0 \\ 0 & A(1|1) \end{bmatrix},$$

and it follows that $AB = BA$.   □

We now turn to Laplacians. Let G be a graph with $V(G) = \{1, \ldots, n\}$. Let i, j be nonadjacent vertices of G, and let H be the graph obtained from G by adding the edge $\{i, j\}$. Then $L(H) = L(G) + e_{ij}e'_{ij}$. Thus, if $\lambda_1 \geq \cdots \geq \lambda_n = 0$ are the eigenvalues of $L(G)$, and $\mu_1 \geq \cdots \geq \mu_n = 0$ are the eigenvalues of $L(H)$, then

$$\mu_1 \geq \lambda_1 \geq \mu_2 \geq \cdots \geq \lambda_{n-1} \geq \mu_n \geq \lambda_n. \tag{11.11}$$

Suppose G is Laplacian integral. Since trace $L(H) = $ trace $L(G) + 2$, then in view of (11.11) H will also be Laplacian integral if either

(a) $n - 1$ eigenvalues of $L(G)$ and $L(H)$ coincide and one eigenvalue of $L(G)$ increases by 2, or
(b) $n - 2$ eigenvalues of $L(G)$ and $L(H)$ coincide, and two eigenvalues of $L(G)$ increase by 1.

We say that spectral integral variation occurs in 1 or 2 places according as (a) or (b) holds, respectively. The case (a) is characterized in the next result. We denote by $N(i)$ the neighbourhood, that is the set of vertices adjacent to, the vertex i.

Theorem 11.13 *Let G be a Laplacian integral graph with $V(G) = \{1, \ldots, n\}$. Let i, j be nonadjacent vertices of G, and H be the graph obtained from G by adding the edge (i, j). Then $n - 1$ eigenvalues of $L(G)$ and $L(H)$ coincide if and only if $N(i) = N(j)$.*

Proof As observed earlier, $L(H) = L(G) + e_{ij}e'_{ij}$. By Theorem 11.12, $n - 1$ eigenvalues of $L(G)$ and $L(H)$ coincide if and only if $L(G)e_{ij}e'_{ij} = e_{ij}e'_{ij}L(G)$. It is easy to see that this condition is equivalent to $N(i) = N(j)$.   □

Corollary 11.14 *Let G be a graph with $V(G) = \{1, \ldots, n\}$. Let i, j be nonadjacent vertices of G such that $N(i) = N(j)$, and let H be the graph obtained from G by adding the edge (i, j). Then G is Laplacian integral if and only if H is Laplacian integral.*

Exercises

1. Let G be a connected graph with $V(G) = \{1, \ldots, n\}$. Let $d_1 \geq \cdots \geq d_n$ be the degree sequence of G. Show that G is threshold if and only if any sequence that majorizes, but does not equal, $d_1, \ldots, d_n$ is not the degree sequence of a graph.
2. Show that for $n \geq 2$ the number of nonisomorphic threshold graphs on n vertices is 2^{n-1}.
3. Prove Theorem 11.9.
4. Show that a cograph is Laplacian integral.

5. Show that if a graph contains P_4, the path on 4 vertices, as an induced subgraph, then it is not a cograph.

6. Let G be a graph with $V(G) = \{1, \ldots, n\}$. The graph G is called a *split graph* if there exists a partition $V(G) = V_1 \cup V_2$ such that the graph induced by V_1 is complete, the graph induced by V_2 has no edge and every vertex in V_1 is adjacent to every vertex in V_2. Find the Laplacian eigenvalues of a split graph. Hence, find the number of spanning trees in $K_m \setminus G$, where $m \geq n$.

7. Let X_1, Y_1, X_2, Y_2 be disjoint sets with $|X_1| = |X_2| = |Y_1| - 1 = |Y_2| - 1 = r$, and let

$$Y_1 = \{a_1, \ldots, a_{r+1}\}, \quad Y_2 = \{b_1, \ldots, b_{r+1}\}.$$

Let G be the graph with vertex set $X_1 \cup Y_1 \cup X_2 \cup Y_2$ and with the edge set defined as follows. Every vertex in X_i is adjacent to every vertex in Y_i, $i = 1, 2$, and a_j is adjacent to b_j, $j = 1, \ldots r + 1$. Show that G is not a cograph but it is Laplacian integral.

8. Let $G \times H$ denote the Cartesian product of the graphs G and H. Show that $K_n \times K_2$ is not a cograph but it is Laplacian integral.

Marshall and Olkin [MO79] is the classical reference on majorization. An encyclopedic reference on threshold graphs is Mahadev and Peled [MP95]. Sections 10.2 and 10.3 are based on Merris [M94] and Wasin So [S99], respectively. A construction of an infinite family of Laplacian integreal graphs that are not cographs is given in [GM08]. A conjecture of Grone and Merris [GM94] asserts that the Laplacian eigenvalues of any graph are majorized by the conjugate of its degree sequence. The conjecture has been settled by Hua Bai [Bai11].

References and Further Reading

[Bai11] Bai, H.: The Grone-Merris conjecture. Trans. Amer. Math. Soc. **363**(8), 4463–4474 (2011)

[GM94] Grone, R., Merris, R.: The Laplacian spectrum of a graph II. SIAM J. Discrete Math. **7**(2), 221–229 (1994)

[GM08] Grone, R., Merris, R.: Indecomposable Laplacian integral graphs. Linear Algebra Appl. **428**, 1565–1570 (2008)

[MP95] Mahadev, N.V.R., Peled, U.N.: Threshold Graphs and Related Topics, Annals of Discrete Mathematics, 54. North-Holland Publishing Co., Amsterdam (1995)

[MO79] Marshall, A.W., Olkin, I.: Inequalities: Theory of Majorization and Its Applications, Mathematics in Science and Engineering, 143. Academic Press, New York (1979)

[M94] Merris, R.: Degree maximal graphs are Laplacian integral. Linear Algebra Appl. **199**, 381–389 (1994)

[S99] So, W.: Rank one perturbation and its application to the Laplacian spectrum of a graph. Linear and Multilinear Algebra **46**, 193–198 (1999)

Chapter 12
Positive Definite Completion Problem

Several problems in mathematics can be viewed as completion problems. Matrix theory is particularly rich in such problems. Such problems nicely blend graph theoretic notions with matrix theory. In this chapter we consider one particular completion problem, the positive definite completion problem, in detail.

12.1 Nonsingular Completion

We illustrate the idea of matrix completion problems by a simple example. We first introduce some terminology. Let G be a bipartite graph with the bipartition (R, C), where $R = (R_1, \ldots, R_n)$ and $C = (C_1, \ldots, C_n)$. A G-partial $n \times n$ matrix is a matrix in which a_{ij} is specified if and only if R_i is adjacent to C_j. By a completion of a G-partial matrix we mean a specification of all the unspecified entries in the matrix. The graph G is called *shape nonsingular completable* if any G-partial matrix admits a nonsingular completion.

Example 12.1 Consider the graph G and the G-partial matrix A:

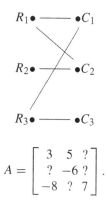

$$A = \begin{bmatrix} 3 & 5 & ? \\ ? & -6 & ? \\ -8 & ? & 7 \end{bmatrix}.$$

© Springer-Verlag London 2014
R.B. Bapat, *Graphs and Matrices*, Universitext,
DOI 10.1007/978-1-4471-6569-9_12

It is easy to see that we can fill up the unspecified entries of A (indicated by the question marks) so that the resulting matrix is nonsingular. In fact any G-partial matrix can be completed to a nonsingular matrix, and hence G is nonsingular completable.

If G is a bipartite graph with the bipartition (R, C) then G^c will denote the bipartite complement of G. Thus, R_i and C_j are adjacent in G^c if and only if they are not adjacent in G. The characterization of nonsingular completable graphs is stated in the next result.

Theorem 12.2 *Let G be a bipartite graph with bipartition (R, C), where*

$$R = (R_1, \ldots, R_n), \quad C = (C_1, \ldots, C_n).$$

Then G is nonsingular completable if and only if G^c has a perfect matching.

Proof First suppose that G^c has a perfect matching, and suppose it is given by the edges $(R_i, C_{\sigma(i)})$, $i = 1, \ldots, n$, where σ is a permutation. Let A be a G-partial matrix. Consider the matrix $A(x)$ obtained by letting the $(R_i, C_{\sigma(i)})$-entry of A be x, $i = 1, \ldots, n$, and specifying the remaining unspecified entries as zero. Then $\det A(x)$ is a polynomial in x of degree n, in which the leading term is $\pm x^n$. Thus, for some value of x, $\det A(x)$ is nonzero and hence $A(x)$ is nonsingular. Therefore, G is nonsingular completable.

Conversely, suppose G^c has no perfect matching. Then by the König–Egervary theorem, G^c has a vertex cover of size less than n. Without loss of generality, let the vertices $R_1, \ldots, R_k$ and $C_1, \ldots, C_s$ form a vertex cover of G^c, where $k + s < n$. Let A be the $n \times n$ G-partial matrix in which $a_{ij} = 0$ whenever R_i is adjacent to C_j in G. Then the submatrix of A formed by the rows $k + 1, \ldots, n$ and the columns $s + 1, \ldots, n$ is zero. Let B be an arbitrary completion of A. Then, since $k + s < n$, it can be seen by Laplace expansion along the first k rows, that $\det B = 0$. Thus, G is not nonsingular completable and the proof is complete.   □

Note that Theorem 12.2 by itself is easy to prove and not a profound result. However, it points to a fertile area of matrix completion problems one may consider. A particularly elegant matrix completion problem is considered in the following sections.

12.2 Chordal Graphs

The class of chordal graphs will be relevant in connection with the positive definite completion problem. Chordal graphs admit many equivalent definitions and arise in several areas. A connected graph G is said to be *chordal* (or *triangulated*) if it does not have C_k, the cycle on k vertices, $k \geq 4$, as an induced subgraph. Equivalently, G is chordal if any C_k, $k \geq 4$, in the graph has a "chord", that is, an edge joining two intermediate vertices in the cycle.

Examples of chordal graphs include K_n, trees and threshold graphs. The following graph, denoted T_n, is also chordal:

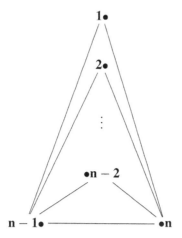

Let G be a graph with $V(G) = \{1, \ldots, n\}$. An ordering $i_1, \ldots, i_n$ of $1, \ldots, n$ is called a *perfect elimination ordering* if, for $j = 1, \ldots, n-1$, the subgraph induced by $\{i_k : k > j, i_k \sim i_j\}$ is a clique (a complete graph).

The following characterization of chordal graphs is well known. We omit the proof.

Theorem 12.3 *A graph is chordal if and only if its vertices admit a perfect elimination ordering.*

As an example, the following graph is chordal and a perfect elimination ordering is given by $1, 2, \ldots, 8$:

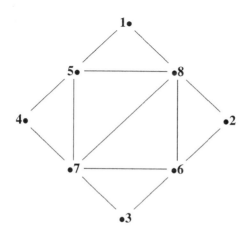

A clique in a graph is said to be *shape maximal* if it is not properly contained in another clique. We now obtain some results concerning chordal graphs that will be used.

Lemma 12.4 *Let G be a chordal graph with $V(G) = \{1, \ldots, n\}$. Let $e = \{i, j\}$ be an edge of G such that $G \setminus \{e\}$ is also chordal. Then i, j, together with all the vertices adjacent to both i and j, form a maximal clique.*

Proof Let K be the subgraph induced by the vertices i, j and all the vertices adjacent to both i and j. If u, v are distinct vertices adjacent to both i and j then we claim that $u \sim v$. Otherwise, i, j, u, v would induce a C_4 in $G \setminus \{e\}$, contradicting the fact that $G \setminus \{e\}$ is chordal. Thus, the claim is proved. It follows that K is a clique and it contains i and j. Furthermore, K is maximal in the sense that there is no clique K' that properly contains K.   □

Lemma 12.5 *Let $G \neq K_n$ be a chordal graph with $V(G) = \{1, \ldots, n\}$. Then there exist $i, j \in V(G)$ such that i is not adjacent to j, and the graph $H = G + e$ obtained by adding the edge $e = \{i, j\}$ to G is chordal.*

Proof We assume, without loss of generality, that $1, 2, \ldots, n$ is a perfect elimination ordering of $V(G)$. Let i be the largest integer with the property that the subgraph induced by $\{i, i + 1, \ldots, n\}$ is not a clique. The existence of i is guaranteed since $G \neq K_n$. Then there exists $j > i, j \not\sim i$. Let $e = \{i, j\}$, and let $H = G + e$. Then $1, 2, \ldots, n$ is a perfect elimination ordering of H as well, and hence H is chordal.   □

12.3 Positive Definite Completion

A partial symmetric $n \times n$ matrix is an $n \times n$ matrix in which some entries are specified and some are unspecified, such that for $i \neq j$, if the (i, j)-entry is specified, then so is the (j, i)-entry, and it is equal to the (i, j)-entry. We also assume that the diagonal entries are all specified. A partial positive definite matrix is a partial symmetric matrix in which any principal submatrix that is completely specified has a positive determinant. A partial positive semidefinite matrix is a partial symmetric matrix in which any principal submatrix that is completely specified has a nonnegative determinant.

Let A be a partial symmetric $n \times n$ matrix. The specification graph G_A associated with A is defined as follows. The vertex set of G_A is $\{1, \ldots, n\}$. For $i \neq j$ the vertices i and j are adjacent if and only if a_{ij} (and hence a_{ji}) is specified. We also say that A has specification graph G_A or that G_A is the specification graph of A. (The word "specificatio" is used since normally the graph associated with a matrix uses the zero-nonzero structure of the matrix.)

Example 12.6 Consider the following matrix A. The unspecified entries are indicated by question marks.

$$A = \begin{bmatrix} 2 & 1 & ? & -1 & 0 \\ 1 & 3 & ? & 1 & ? \\ ? & ? & 2 & 0 & ? \\ -1 & 1 & 0 & 2 & 0 \\ 0 & ? & ? & 0 & 1 \end{bmatrix}.$$

It can be checked that A is partial positive definite.

Let G be a graph with $V(G) = \{1, \ldots, n\}$. We say that G is positive definite completable if any partial positive definite matrix A with the specification graph G is completable to a positive definite matrix. Similarly, we say that G is positive semidefinite completable if any partial positive semidefinite matrix A with the specification graph G is completable to a positive semidefinite matrix.

Lemma 12.7 *A graph is positive definite completable if and only if it is positive semidefinite completable.*

Proof First suppose that the graph G is positive semidefinite completable and let A be a partial positive definite matrix with the specification graph G. There exists $\varepsilon > 0$ such that $B = A - \varepsilon I$ is partial positive definite. Since the specification graph of B is G as well, B is completable to a positive semidefinite matrix, say $\tilde{B}$. Then $\tilde{A} = \tilde{B} + \varepsilon I$ is a positive definite completion of A. Therefore, G is positive definite completable.

Conversely, suppose G is positive definite completable. Let A be a partial positive semidefinite matrix. For any positive integer k, let $B_k = A + \frac{1}{k}I$. Then B_k is a partial positive definite matrix with the specification graph G and therefore B_k is completable to a positive definite matrix, say $\tilde{B}_k$. Note that the off-diagonal entries of a positive semidefinite matrix are bounded in modulus by the largest diagonal entry. Since the diagonal entries of $\tilde{B}_k$ are bounded by $\max_i\{a_{ii} + 1\}$, the matrices $\tilde{B}_k$, $k = 1, 2, \ldots$ (or a subsequence thereof) converge to a matrix, say B. Then B is a positive semidefinite completion of A. Hence, G is positive semidefinite completable and the proof is complete.   □

Lemma 12.8 C_4 *is not positive definite completable.*

Proof By Lemma 12.7 it will be sufficient to show that C_4 is not positive semidefinite completable. Let

$$B = \begin{bmatrix} 1 & 1 & x \\ 1 & 1 & 1 \\ x & 1 & 1 \end{bmatrix}.$$

Then $\det B = -(1 - x)^2$. It follows that B is positive semidefinite if and only if $x = 1$. Consider the partial positive semidefinite matrix A with the specification graph C_4:

$$A = \begin{bmatrix} 1 & 1 & ? & 0 \\ 1 & 1 & 1 & ? \\ ? & 1 & 1 & 1 \\ 0 & ? & 1 & 1 \end{bmatrix}.$$

It follows by the preceding observation that in order to complete A to a positive semidefinite matrix, the $(1, 3), (2, 4)$ entries (and hence the $(3, 1), (4, 2)$ entries) must be set equal to 1. But then, since the $(1, 4)$-entry is 0, a positive semidefinite completion is not possible. Therefore, C_4 is not positive semidefinite completable.   □

The Jacobi identity for determinants asserts that if A is a nonsingular $n \times n$ matrix, and if $B = A^{-1}$, then for any nonempty, proper subsets S, T of $\{1, \ldots, n\}$, with $|S| = |T|$,

$$\det B[S|T] = \frac{\det A(T|S)}{\det A}.$$

The identity can be proved using the formula for the inverse of a partitioned matrix, and the Schur complement formula for the determinant.

Lemma 12.9 *Let A be an $n \times n$ matrix and let $i, j \in \{1, \ldots, n\}, i \neq j$. Then*

$$\det A(i|i) \det A(j|j) - \det A(i|j) \det A(j|i) = (\det A)(\det A(i, j|i, j)).$$

Proof It will be sufficient to prove the result when A is nonsingular, as the general case can be derived by a continuity argument. So suppose A is nonsingular, and let $B = A^{-1}$. By the Jacobi identity for the determinant,

$$\det B[i, j|i, j] = \frac{\det A(i, j|i, j)}{\det A}. \tag{12.1}$$

Note that

$$B[i, j|i, j] = \frac{1}{\det A} \begin{bmatrix} \det A(i|i) & (-1)^{i+j} \det A(i|j) \\ (-1)^{i+j} \det A(i|j) & \det A(j|j) \end{bmatrix},$$

and therefore

$$\det B[i, j|i, j] = \left(\frac{1}{\det A}\right)^2 (\det A(i|i) \det A(j|j) - \det A(i|j) \det A(j|i)). \tag{12.2}$$

The result follows from (12.1) and (12.2).   □

Lemma 12.10 *Let $i, j \in \{1, \ldots, n\}, i \neq j$, and let $e = \{i, j\}$ be an edge of K_n. The graph $K_n \setminus \{e\}$ is positive definite completable.*

Proof We assume, without loss of generality, that $e = \{1, n\}$. Let A be an $n \times n$ matrix with the specification graph $K_n \setminus \{e\}$, and suppose A is partial positive definite. Specify the $(1, n)$-entry of A as x. We continue to denote the resulting matrix as A for convenience. Since A is symmetric, by Lemma 12.9,

$$\det A(1|1) \det A(n|n) - (\det A(1|n))^2 = (\det A)(\det A(1, n|1, n)). \tag{12.3}$$

Note that

$$\det A(1|n) = (-1)^{n+1} x \det A(1, n|1, n) + \alpha \qquad (12.4)$$

for some α. Since A is partial positive definite, $\det A(1, n|1, n) > 0$. Let

$$x_0 = (-1)^n \frac{\alpha}{\det A(1, n|1, n)}.$$

Specify the $(1, n)$-entry of A as x_0. We continue to denote the resulting matrix by A. By (12.4) $\det A(1|n) = 0$ and hence by (12.3),

$$\det A = \frac{\det A(1|1) \det A(n|n)}{\det A(1, n|1, n)} > 0. \qquad (12.5)$$

For $k = 1, \ldots, n - 1$, the leading principal minor of A formed by the rows and the columns $\{1, \ldots, k\}$ is positive since A is partial positive definite. As observed in (12.5), $\det A > 0$ and hence A is positive definite. Thus, any partial positive definite matrix A with the specification graph $K_n \backslash \{e\}$ admits a positive definite completion and hence $K_n \backslash \{e\}$ is positive definite completable. $\qquad \square$

We are now in a position to present a characterization of positive definite completable matrices.

Theorem 12.11 *Let G be a graph with vertices $\{1, \ldots, n\}$. Then G is positive definite completable if and only if G is chordal.*

Proof First suppose G is chordal. If $G = K_n$ then clearly G is positive definite completable. So suppose $G \neq K_n$. By Lemma 12.5 there exist $i, j \in V(G)$ such that i is not adjacent to j, and the graph $H = G + e$ obtained by adding the edge $e = \{i, j\}$ to G is chordal. By Lemma 12.4 there exists a maximal clique K in H containing i, j and and all the vertices adjacent to both i and j. Let A be a partial positive definite matrix with the specification graph G.

Let B be the principal submatrix of A, indexed by the rows and the columns in $V(K)$, the set of vertices of K. Note that B is partial positive definite, and its specification graph is a complete graph, with a single missing edge. By Lemma 12.10 we can complete B to a positive definite matrix. Thus, we can specify the (i, j)-entry (and the (j, i)-entry) of A so that the resulting matrix, say A_1, is partial positive definite. The specification graph of A_1 is H, which is chordal. We may continue this process until we obtain a positive definite completion of A.

Conversely, suppose G is not chordal. Then G has C_4 as an induced subgraph. By Lemma 12.8 C_4 is not positive definite completable and hence G is not positive definite completable. This completes the proof. $\qquad \square$

Exercises

1. Let m, n be positive integers and let $1 \leq k \leq \min\{m, n\}$. The specification graph of a partial $m \times n$ matrix is a bipartite graph, with bipartite sets of cardinality m and n defined in the usual way. Call a graph G rank k completable if any partial matrix with the specification graph G can be completed to a matrix of rank at least k. Characterize rank k completable graphs.
2. Let G be a graph with $V(G) = \{1, \ldots, n\}$. Recall that the graph G is called a split graph if there exists a partition $V(G) = V_1 \cup V_2$ such that the graph induced by V_1 is complete and the graph induced by V_2 has no edge. Show that if G is a split graph, then both G and G^c are chordal.
3. Give an example of a graph G that is not chordal and a partial positive definite matrix A with specification graph G, which admits a positive definite completion.
4. Give an example to show that the positive definite completion of a partial positive definite matrix need not be unique.
5. Show that the following matrix can be reduced to a diagonal matrix by elementary row and column operations so that the zero entries in the matrix are never made nonzero:

$$\begin{bmatrix} 6 & 0 & 1 & 0 & 0 & 0 & 0 \\ 0 & 6 & 1 & 0 & 0 & 0 & 1 \\ 1 & 1 & 6 & 1 & 1 & 1 & 1 \\ 0 & 0 & 1 & 6 & 0 & 1 & 1 \\ 0 & 0 & 1 & 0 & 6 & 0 & 0 \\ 0 & 0 & 1 & 1 & 0 & 6 & 1 \\ 0 & 1 & 1 & 1 & 0 & 1 & 6 \end{bmatrix}$$

6. Let A be an $n \times n$ orthogonal matrix and let S and T be nonempty, proper subsets of $\{1, \ldots, n\}$, with $|S| = |T|$. Show that

$$\det A[S|T] = \pm \det A(S|T).$$

Theorem 12.11 was proved in [GJSW84]. Our exposition is partly based on [BS03]. Chordal graphs are discussed in greater detail in [GOL80].

References and Further Reading

[BS03]   Berman, A., Shaked-Monderer, N.: Completely Positive Matrices. World Scientific, Singapore (2003)
[GOL80]   Golumbic, M.: Algorithmic Graph Theory and Perfect Graphs. Academic Press, New York (1980)
[GJSW84] Grone, R., Johnson, C.R., Sá, E.M., Wolkowitz, H.: Positive definite completions of partial hermitian matrices. Linear Algebra Appl. **58**, 109–124 (1984)

Chapter 13
Matrix Games Based on Graphs

In this chapter we consider two-person zero-sum games, or matrix games, in which the pure strategies of the players are the vertices, or the edges of a graph, and the payoff is determined by the incidence structure. We identify some cases where the value and the optimal strategies can be explicitly determined. We begin with a brief overview of the theory of matrix games.

13.1 Matrix Games

Suppose there are two players, I and II. Player I has m pure strategies $\{1, \ldots, m\}$, while Player II has n strategies $\{1, \ldots, n\}$. If Player I selects the strategy i and Player II selects the strategy j, then Player I receives the amount a_{ij} from Player II, $i = 1, \ldots, m; j = 1, \ldots, n$. The $m \times n$ matrix $A = [a_{ij}]$ is called the payoff matrix of this game. Since the gain of Player I is the loss of Player II, a matrix game is also known as a two-person zero-sum game.

The strategy set naturally extends to mixed strategies. A mixed strategy for a player is a probability distribution over the set of pure strategies. Let $\mathscr{P}_k$ denote the set of probability vectors of order $k \times 1$. Thus,

$$\mathscr{P}_k = \left\{ x \in \mathbb{R}^k : x_i \geq 0, \quad i = 1, \ldots, k; \quad \sum_{i=1}^{k} x_i = 1 \right\}.$$

If Player I selects $x \in \mathscr{P}_m$ and Player II selects $y \in \mathscr{P}_n$, then the payoff to Player I from Player II is taken to be the expected value of the payoff, which equals $x'Ay = \sum_{i=1}^{m} \sum_{j=1}^{n} a_{ij} x_i y_j$.

A pair of strategies $(x, y) \in \mathscr{P}_m \times \mathscr{P}_n$ are said to be in equilibrium, or they are a pair of optimal strategies, if x is a best response of Player I if Player II chooses y; and y is a best response of Player II if Player I chooses x. Equivalently, $x \in \mathscr{P}_m$ is optimal for Player I if it maximizes

© Springer-Verlag London 2014
R.B. Bapat, *Graphs and Matrices*, Universitext,
DOI 10.1007/978-1-4471-6569-9_13

$$\min_{z \in \mathscr{P}_n} \{x'Az\},$$

while $y \in \mathscr{P}_n$ is optimal for Player II if it minimizes

$$\max_{z \in \mathscr{P}_m} \{z'Ay\}.$$

We now state the well-known minimax theorem of von Neumann.

Theorem 13.1 *Let Players I and II have m and n pure strategies, respectively, and let A be the $m \times n$ payoff matrix. Then there exist optimal strategies $x \in \mathscr{P}_m$ and $y \in \mathscr{P}_n$. Furthermore, there is a unique real number v (known as the value of the game) such that*

$$x'A \geq v\mathbf{1}', \quad Ay \leq v\mathbf{1}.$$

If A is an $m \times n$ matrix we denote the value of the matrix game A by $v(A)$.

Corollary 13.2 *Let A be an $m \times n$ matrix. Let $p \in \mathscr{P}_m$, $q \in \mathscr{P}_n$, and let α be a real number, such that*

$$p'A \geq \alpha\mathbf{1}', \quad Aq \leq \alpha\mathbf{1}. \tag{13.1}$$

Then $v(A) = \alpha$, and p and q are optimal strategies for Players I and II, respectively.

Proof Let x and y be optimal strategies for Players I and II, respectively, as guaranteed by Theorem 13.1, so that

$$x'A \geq v(A)\mathbf{1}', \quad Ay \leq v(A)\mathbf{1}. \tag{13.2}$$

It follows from (13.1) to (13.2) that

$$p'Ay \geq \alpha, \quad p'Ay \leq v(A) \tag{13.3}$$

and

$$x'Aq \geq v(A), \quad x'Aq \leq \alpha. \tag{13.4}$$

Using (13.3) and (13.4) we conclude that $\alpha = v(A)$. Then by (13.1), p and q are optimal for Players I and II, respectively. $\square$

Example 13.3 Consider the two payoff matrices

$$A = \begin{bmatrix} 3 & 2 \\ 4 & 1 \end{bmatrix}, \quad B = \begin{bmatrix} 3 & 5 \\ 6 & 4 \end{bmatrix}.$$

It can be verified that for the matrix game A, there are pure optimal strategies for both the players, strategy 1 for Player I and strategy 2 for Player II. The value of the game is 2. In the case of the matrix game B, there are no pure optimal strategies.

If $x = [\frac{1}{2}, \frac{1}{2}]'$, $y = [\frac{1}{4}, \frac{3}{4}]'$, then x and y are optimal for the two players, respectively. The value of the game is $\frac{9}{2}$.

Let A be an $m \times n$ matrix. The set of optimal strategies of Player I and Player II will be denoted by $Opt_I(A)$ and $Opt_{II}(A)$, respectively. The dimension of $Opt_I(A)$ is defined as the dimension of the vector space spanned by $Opt_I(A)$, minus 1. The dimension of $Opt_{II}(A)$ is defined similarly. Note that a player has a unique optimal strategy if and only if the dimension of the set of its optimal strategies is zero. A pure strategy is called *essential* if it is used with positive probability in some optimal strategy. Otherwise it is called *inessential*. We now state two classical results, without proof.

Theorem 13.4 *Let A be an $m \times n$ matrix. Let $S \subset \{1, \ldots, m\}$, $T \subset \{1, \ldots, n\}$ be the sets of essential strategies of Players I and II, respectively. Let $B = A[S|T]$. Then*

$$\dim(Opt_I(A)) = \text{nullity}(B') - 1 = |S| - \text{rank } B - 1$$

and

$$\dim(Opt_{II}(A)) = \text{nullity}(B) - 1 = |T| - \text{rank } B - 1.$$

Theorem 13.5 *Let A be an $m \times n$ matrix. Let f_1 and f_2 be the number of essential strategies of Players I and II, respectively. Then*

$$f_1 - \dim(Opt_I(A)) = f_2 - \dim(Opt_{II}(A)).$$

13.2 Vertex Selection Games

Let G be a directed graph with $V(G) = \{1, \ldots, n\}$. In the vertex selection game, Players I and II independently select a vertex of G. If Player I selects i and Player II selects j, then Player I receives 1 or -1 from Player II according as there is an edge from i to j or from j to i, respectively. If $i = j$ or if i and j are not adjacent then Player I receives nothing from Player II. The payoff matrix of the vertex selection game is the $n \times n$ matrix A defined as follows. The rows and the columns of A are indexed by $V(G)$. If $i = j$ or if i and j are not adjacent then $a_{ij} = 0$. Otherwise $a_{ij} = 1$ or -1 according as there is an edge from i to j or from j to i, respectively. We will refer to A as the *skew matrix* of the graph G. This terminology is justified since A is skew-symmetric. We assume that the graph G has at least one edge, although this fact may not be stated explicitly.

If a matrix is skew-symmetric then the associated game is symmetric with respect to the two players. A special property enjoyed by such matrix games is given in the next result.

Lemma 13.6 *Let A be an $n \times n$ skew-symmetric matrix. Then $v(A) = 0$. Furthermore, Players I and II have identical optimal strategy sets.*

Proof Let x and y be optimal strategies for Players I and II, respectively. Then

$$x'A \geq v(A)\mathbf{1}', \quad Ay \leq v(A)\mathbf{1}. \tag{13.5}$$

Since $A' = -A$, it follows from (13.5) that

$$Ax \leq -v(A)\mathbf{1}, \quad y'A \geq -v(A)\mathbf{1}'. \tag{13.6}$$

Following the proof of Corollary 13.2, we obtain from (13.5) and (13.6) that $v(A) = -v(A)$, and hence $v(A) = 0$. It is evident from (13.6) that x is optimal for Player II and y is optimal for Player I. Therefore, Players I and II have identical optimal strategy sets. $\quad\square$

The vertex selection game associated with the graph G is the matrix game with payoff matrix A, which is the skew matrix of G. Since the skew matrix is skew-symmetric, we conclude from Lemma 13.6 that the vertex selection game has value zero and the two players have identical strategy sets. We will now be concerned with some properties of the optimal strategies in vertex selection games. We begin with some preliminary observations. Recall that a vertex of a directed graph is called a *source* if its indegree is zero, while a vertex is called a *sink* if its outdegree is zero.

Lemma 13.7 *Let G be a directed graph with $V(G) = \{1, \ldots, n\}$, and let A be the skew matrix of G. The pure strategy i is optimal (for either player) if and only if the vertex i is a source.*

Proof Let u_i be the $n \times 1$ unit vector, that is, the vector with the ith coordinate equal to 1 and the remaining coordinates equal to 0. The pure strategy i is represented by the vector u_i. As observed earlier, $v(A) = 0$. Thus, the strategy u_i is optimal for Player I if and only if $u_i'A \geq 0$, or, equivalently, if the ith row of A has no negative element. Clearly, this is equivalent to vertex i having indegree 0. $\quad\square$

Example 13.8 Consider the directed path on 5 vertices,

$$\bullet 1 \longrightarrow \bullet 2 \longrightarrow \bullet 3 \longrightarrow \bullet 4 \longrightarrow \bullet 5$$

with the skew matrix

$$\begin{bmatrix} 0 & 1 & 0 & 0 & 0 \\ -1 & 0 & 1 & 0 & 0 \\ 0 & -1 & 0 & 1 & 0 \\ 0 & 0 & -1 & 0 & 1 \\ 0 & 0 & 0 & -1 & 0 \end{bmatrix}.$$

The vertex 1 has indegree zero and the pure strategy 1 represented by the vector $[1, 0, 0, 0, 0]'$ is optimal. It may be noted that the strategy $[\frac{1}{3}, 0, \frac{1}{3}, 0, \frac{1}{3}]'$ is also optimal, and this strategy selects the vertex 5 with positive probability, even though this vertex is a sink.

13.3 Tournament Games

We first prove a preliminary result.

Lemma 13.9 *Let A be an $m \times n$ matrix and let x and y be optimal strategies for Players I and II respectively. Then $x_i > 0$ implies $(Ay)_i = v(A)$, and $y_j > 0$ implies $(x'A)_j = v(A)$.*

Proof Since x and y are optimal for Players I and II, respectively,

$$x'A \geq v(A)\mathbf{1}', \quad Ay \leq v(A)\mathbf{1}.$$

From these inequalities we easily derive that $x'Ay = v(A)$. If $x_i > 0$ and $(Ay)_i < v(A)$ for some i, then it would lead to $x'Ay < v(A)$, which is a contradiction. Hence, $x_i > 0$ implies $(Ay)_i = v(A)$. The second part is proved similarly. $\square$

Corollary 13.10 *Let A be an $n \times n$ skew-symmetric matrix and let x and y be optimal strategies for Players I and II, respectively. Then $y_i > 0$ implies $(Ax)_i = 0$.*

Proof By Lemma 13.6, $v(A) = 0$. Now the result follows from Lemma 13.9. $\square$

A *tournament* is defined as a directed graph obtained by the orienting of each edge of a complete graph. A tournament with n vertices may represent the results of a competition among n players in which any two players play against each other and there are no draws. We now consider vertex selection games corresponding to tournaments. The well-known "scissors, paper and stone" game is the same as the vertex selection game corresponding to the directed 3-cycle, or a tournament with 3 vertices. We define a tournament game as the vertex selection game corresponding to a tournament; such a game provides a generalization of the scissors, paper and stone game.

Lemma 13.11 *Let T be a tournament with $V(T) = \{1, \ldots, n\}$ and let A be the skew matrix of T. Then the rank of A is n if n is even and $n - 1$ if n is odd.*

Proof Replace each off-diagonal entry of A by 1 and let B be the resulting matrix. First observe that det A and det B are either both even or are both odd. By Theorem 3.4 the eigenvalues of B are $n - 1$ and -1 with multiplicity $n - 1$. Therefore, det $B = (n - 1)(-1)^{n-1}$. Thus, if n is even then det B, and hence det A is odd. Therefore, det A is nonzero and the rank of A is n. If n is odd, we may apply the same argument to a subtournament of T, consisting of $n - 1$ vertices, and deduce that the rank of A is at least $n - 1$. Note that since $A' = -A$ then det $A' = (-1)^n$ det A, and since n is odd it follows that det $A = 0$. Thus, A is singular and its rank must be $n - 1$. $\square$

Corollary 13.12 *Let T be a tournament with $V(T) = \{1, \ldots, n\}$, and suppose there is an optimal strategy x with $x > 0$ in the corresponding tournament game. Then n is odd.*

Proof Let A be the skew matrix of T. By Corollary 13.10, $x_i > 0$ implies $(Ax)_i = 0$. Since $x_i > 0$ for each i, $Ax = 0$ and hence rank $A < n$. It follows by Lemma 13.11 that n is odd. $\square$

We now prove the main result concerning optimal strategies in tournament games.

Theorem 13.13 *Let T be a tournament with $V(T) = \{1, \ldots, n\}$. Then there is a unique optimal strategy for the corresponding tournament game.*

Proof Let A be the skew matrix of T. Let p and q be optimal strategies for the tournament game corresponding to T. Let

$$S = \{i : 1 \leq i \leq n, \ p_i > 0 \text{ or } q_i > 0\}.$$

Let $B = A[S|S]$ and let p_S and q_S be the subvectors of p and q corresponding to S, respectively. Now using Lemma 13.6 and Corollary 13.10, it follows that $Bp_S = Bq_S = 0$. Since $p_S \neq 0$, and since by Lemma 13.11 the nullity of B is at most 1, then $q_S = \alpha p_S$ for some $\alpha \neq 0$. Since $\mathbf{1}'p_S = \mathbf{1}'q_S = 1$, it follows that $p_S = q_S$, and hence $p = q$. Therefore, A has a unique optimal strategy. $\square$

Corollary 13.14 *Let G be a graph with $V(G) = \{1, \ldots, n\}$. Then $G = K_n$ if and only if the vertex selection game corresponding to any orientation of G has a unique optimal strategy.*

Proof If $G = K_n$ then by Theorem 13.13 the vertex selection game corresponding to any orientation of G has a unique optimal strategy. For the converse, suppose $G \neq K_n$, and, without loss of generality, suppose vertices 1 and 2 are not adjacent. We may endow G with an orientation in which both 1 and 2 are source vertices. By Lemma 13.7, in the corresponding vertex selection game the pure strategy 1 as well as the pure strategy 2 are both optimal. Thus, there is an orientation of G such that the corresponding vertex selection game does not have a unique optimal strategy, and the proof is complete. $\square$

We now indicate another approach to Theorem 13.13. Let G be a directed graph with $V(G) = \{1, \ldots, n\}$. Let A be the skew matrix of G and consider the corresponding matrix game. Recall that the optimal strategy sets of Players I and II are identical, and hence so are the essential strategies of the two players. Thus, in this case we obtain the following consequence of Theorem 13.4.

Theorem 13.15 *Let G be a directed graph with $V(G) = \{1, \ldots, n\}$. Let A be the skew matrix of G and let $S \subset \{1, \ldots, n\}$ be the set of essential strategies. Let $B = A[S|S]$. Then*

$$\dim(Opt_I(A)) = \dim(Opt_{II}(A)) = \text{nullity}(B) - 1 = |S| - \text{rank } B - 1.$$

Let T be a tournament with $V(T) = \{1, \ldots, n\}$, and let A be the skew matrix of T. Let $S \subset \{1, \ldots, n\}$ be the set of essential strategies and let $B = A[S|S]$. By

Lemma 13.11, the rank of B is either $|S|$ or $|S| - 1$. In view of Theorem 13.15 we see that the rank must be $|S| - 1$, since the dimension cannot be negative. It also follows that the dimension of $Opt_I(A)$ and $Opt_{II}(A)$ is zero, and hence the optimal strategy is unique, leading to another verification of Theorem 13.13.

13.4 Incidence Matrix Games

Let G be a directed graph with $V(G) = \{1, \ldots, n\}$ and $E(G) = \{e_1, \ldots, e_m\}$. Consider the following two-person zero-sum game. The pure strategy sets of Players I and II are $V(G)$ and $E(G)$, respectively. If Player I selects i and Player II selects e_j, then the payoff to Player I from Player II is defined as follows. If i and e_j are not incident then the payoff is zero. If e_j originates at i then the payoff is 1, while if e_j terminates at i then the payoff is -1. Clearly the payoff matrix of this game is the (vertex-edge) incidence matrix Q of G. We refer to this game as the incidence matrix game corresponding to G.

Lemma 13.16 *Let G be a directed graph with $V(G) = \{1, \ldots, n\}$ and $E(G) = \{e_1, \ldots, e_m\}$. Let Q be the $n \times m$ incidence matrix of G. Then $0 \leq v(Q) \leq 1$. Furthermore, $v(Q) = 0$ if G has a directed cycle, and $v(Q) = 1$ if G is the star $K_{1,n-1}$, with the central vertex being a source.*

Proof The strategy $z = \frac{1}{n}\mathbf{1}$ for Player I satisfies $z'Q = 0$. Let y be optimal for Player II so that $Qy \leq v(Q)\mathbf{1}$. Then $v(Q) \geq z'Qy = 0$. Since $q_{ij} \leq 1$ for all i, j, it follows that $v(Q) \leq 1$.

Suppose G has a directed cycle with k vertices. Consider the strategy z for Player II, who chooses each edge of the cycle with probability $\frac{1}{k}$. Then $Qz = 0$. Let x be optimal for Player I, so that $x'Q \geq v(Q)\mathbf{1}'$. Hence, $v(Q) \leq x'Qz = 0$. Since we have shown earlier that $v(Q) \geq 0$, it follows that $v(Q) = 0$.

Now suppose $G = K_{1,n-1}$, and let 1 be the central vertex, which is assumed to be a source. It can be verified that the pure strategy 1 for Player I and any pure strategy for Player II are optimal and $v(Q) = 1$. $\square$

It is evident from Lemma 13.16 that if G has a directed cycle, then the incidence matrix corresponding to G has value 0, and the optimal strategies are easily determined. We now assume that G is acyclic. As usual, let $V(G) = \{1, \ldots, n\}$ and $E(G) = \{e_1, \ldots, e_m\}$. For each $i \in V(G)$ let $P(i)$ denote a path originating at i and having maximum length. Let $\rho(i)$ denote the length (the number of edges) in $P(i)$. If i is a sink then we set $\rho(i) = 0$. For each edge $e_j \in E(G)$, let $\eta(e_j)$ denote the number of vertices i such that e_j is on the path $P(i)$. With this notation we have the following.

Lemma 13.17 $\displaystyle\sum_{i=1}^{n} \rho(i) = \sum_{j=1}^{m} \eta(e_j).$

Proof Let B be the $n \times m$ matrix defined as follows. The rows of B are indexed by $V(G)$, and the columns of B are indexed by $E(G)$. If $i \in V(G)$ and $e_j \in E(G)$ then the (i, j)-entry of B is 1 if $e_j \in P(i)$ and 0, otherwise. Observe that the row sums of B are $\rho(1), \ldots, \rho(n)$ and the column sums of B are $\eta(e_1), \ldots, \eta(e_m)$. Since the sum of the row sums must equal that of the column sums, the result is proved.   □

Theorem 13.18 *Let G be a directed graph with $V(G) = \{1, \ldots, n\}$ and $E(G) = \{e_1, \ldots, e_m\}$. Let Q be the $n \times m$ incidence matrix of G. Let*

$$\sum_{i=1}^{n} \rho(i) = \sum_{j=1}^{m} \eta(e_j) = \frac{1}{\theta}.$$

Then $v(Q) = \theta$. Furthermore, $\theta\rho$ and $\theta\eta$ are optimal strategies for Players I and II, respectively, where ρ is the $n \times 1$ vector with components $\rho(1), \ldots, \rho(n)$ and η is the $m \times 1$ vector with components $\eta(e_1), \ldots, \eta(e_m)$.

Proof First note that by Lemma 13.17,

$$\sum_{i=1}^{n} \rho(i) = \sum_{j=1}^{m} \eta(e_j),$$

and hence θ is well-defined. Fix $j \in \{1, \ldots, m\}$ and suppose the edge e_j is from ℓ to k. We have

$$\theta \sum_{i=1}^{n} q_{ij}\rho(i) = \theta(\rho(\ell) - \rho(k)), \tag{13.7}$$

Note that $\rho(\ell) \geq \rho(k) + 1$ and therefore it follows from (13.7) that

$$\theta \sum_{i=1}^{n} q_{ij}\rho(i) \geq \theta. \tag{13.8}$$

Fix $i \in \{1, \ldots, n\}$ and let

$$U = \{j : e_j \text{ originates at } i\}, \quad W = \{j : e_j \text{ terminates at } i\}.$$

We have

$$\theta \sum_{j=1}^{n} q_{ij}\eta(e_j) = \theta \left(\sum_{j \in U} \eta(e_j) - \sum_{j \in W} \eta(e_j) \right). \tag{13.9}$$

If $U = \phi$, that is, if i is a sink, then the right hand side of (13.9) is clearly nonpositive. Suppose that $U \neq \phi$. Observe that for any vertex $s \neq i$, the path $P(s)$ either contains exactly one edge from U and one edge from W or has no intersection with either U or W. Thus, for any $s \neq i$, the path $P(s)$ either makes a contribution of 1 to both $\sum_{j \in U} \eta(e_j)$ and $\sum_{j \in W} \eta(e_j)$, or does not contribute to either of these terms. Also, the path $P(i)$ makes a contribution of 1 to $\sum_{j \in U} \eta(e_j)$ but none to $\sum_{j \in W} \eta(e_j)$. Thus, if i is not a sink, then

$$\sum_{j \in U} \eta(e_j) - \sum_{j \in W} \eta(e_j) = 1.$$

In view of these observations, we conclude from (13.9) that for $i \in \{1, \ldots, n\}$,

$$\theta \sum_{j=1}^{n} q_{ij} \eta(e_j) \leq \theta. \tag{13.10}$$

The result is proved combining (13.8) and (13.10). $\qquad\square$

Corollary 13.19 *Let G be a directed graph with $V(G) = \{1, \ldots, n\}$ and $E(G) = \{e_1, \ldots, e_m\}$, and let Q be the $n \times m$ incidence matrix of G. Then $v(Q) = 0$ if and only if G has a directed cycle, and $v(Q) = 1$ if and only if G is a star with the central vertex being a source.*

Proof The "if" parts were proved in Lemma 13.16, while the "only if" parts follows from Theorem 13.18. $\qquad\square$

Theorem 13.20 *Let G be a directed graph with $V(G) = \{1, \ldots, n\}$ and $E(G) = \{e_1, \ldots, e_m\}$. Let Q be the $n \times m$ incidence matrix of G. Consider the incidence matrix game corresponding to G. Then Player I has a unique optimal strategy.*

Proof Suppose $\{\phi(i), i \in V(G)\}$ is optimal for Player I. Let $k \in V(G)$ be a sink. Let $y \in \mathscr{P}_m$ be optimal for Player II. If $\phi(k) > 0$ then by Corollary 13.10, we must have

$$\sum_{j=1}^{m} q_{kj} y_j = v(Q). \tag{13.11}$$

Since k is a sink, $q_{kj} \leq 0$, $j = 1, \ldots, m$, whereas by Theorem 13.18 $v(Q) > 0$. This contradicts (13.11) and hence $\phi(k) = 0$.

Let $u \in V(G)$ be a vertex that is not a sink, and let $u = u_0, u_1, \ldots, u_k = w$ be a directed path of maximum length, originating at u. Since ϕ is optimal,

$$\phi(u_i) - \phi(u_{i+1}) \geq v(Q), \quad i = 0, 1, \ldots, k - 1.$$

Thus,

$$\sum_{i=0}^{k-1}(\phi(u_i) - \phi(u_{i+1})) \geq kv(Q),$$

and hence

$$\phi(u) - \phi(w) \geq \rho(u)v(Q).$$

Since w must necessarily be a sink, $\phi(w) = 0$ by our earlier observation, and hence

$$\phi(u) \geq \rho(u)v(Q). \tag{13.12}$$

Thus,

$$1 = \sum_{u \in V(G)} \phi(u) \geq v(Q) \sum_{u \in V(G)} \rho(u) = 1,$$

where the last equality follows from Theorem 13.18. Thus, equality must occur in (13.12), and

$$\phi(u) = \rho(u)v(Q), \quad u \in V(G).$$

Therefore, the strategy of Player I is unique.   □

Example 13.21 Consider the directed, acyclic graph G:

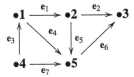

Longest paths emanating from each vertex are given below:

v	$P(v)$
1	e_1, e_5, e_6
2	e_5, e_6
3	ϕ
4	e_3, e_1, e_5, e_6
5	e_6

It can be verified that $\rho(1) = 3$, $\rho(2) = 2$, $\rho(3) = 0$, $\rho(4) = 4$, $\rho(5) = 1$, whereas $\eta(e_1) = 2$, $\eta(e_2) = 0$, $\eta(e_3) = 1$, $\eta(e_4) = 0$, $\eta(e_5) = 3$, $\eta(e_6) = 4$. These, multiplied by $1/10$, are the optimal strategies for Players I and II, respectively, in the incidence matrix game corresponding to G, and the value of the game is $\frac{1}{10}$.

We turn to the optimal strategy space for Player II. Let G be a directed graph with $V(G) = \{1, \ldots, n\}$ and $E(G) = \{e_1, \ldots, e_m\}$. Let Q be the $n \times m$ incidence matrix of G. Consider the incidence matrix game corresponding to G. By Theorem 13.20

the optimal strategy for Player I is unique. By Theorem 13.18 any vertex that not a sink is essential for Player I. Let s be the number of sinks in G and let t be the number of inessential strategies (that is, edges) for Player II. Using the notation of Theorem 13.5, we have $f_1 = n - s$, $f_2 = m - t$. Since $\dim(Opt_I(A)) = 0$, we conclude by Theorem 13.5 that

$$\dim(Opt_{II}(A)) = m - n - t + s.$$

We now consider the 0–1 incidence matrix of an undirected graph and discuss some results for the value of the corresponding matrix game. Let G be a graph with $V(G) = \{1, \ldots, n\}$ and $E(G) = \{e_1, \ldots, e_m\}$. We recall some terminology. A set of edges constitute a matching if no two edges in the set are incident with a common vertex. The maximum cardinality of a matching is called the matching number of G, denoted by $\nu(G)$. A set of vertices of G form a vertex cover if they are collectively incident with all the edges in G. The minimum cardinality of a vertex cover is the vertex covering number of G, denoted by $\tau(G)$.

Lemma 13.22 *Let G be a graph with n vertices and m edges. Let M be the $n \times m$, 0–1 incidence matrix of G. Then*

$$\frac{1}{\tau(G)} \leq \nu(M) \leq \frac{1}{\nu(G)}.$$

Proof Let $\tau(G) = k$, $\nu(G) = \ell$, and suppose, without loss of generality, that the vertices $1, \ldots, k$ form a vertex cover and that the edges $e_1, \ldots, e_\ell$ form a matching. If Player I chooses the vertices $1, \ldots, k$ uniformly with probability $\frac{1}{k}$, then against any pure strategy of Player II, Player I is guaranteed a payoff of at least $\frac{1}{k}$. Similarly, if Player II chooses the edges $e_1, \ldots, e_\ell$ uniformly with probability $\frac{1}{\ell}$, then against any pure strategy of Player I, Player II loses at most $\frac{1}{\ell}$. These two observations together give the result. □

A graph is said to have a *perfect matching* if it has a matching in which the edges are collectively incident with all the vertices. A graph is Hamiltonian if it has a cycle, called a *Hamiltonian cycle*, containing every vertex exactly once.

In the next result we identify some classes of graphs for which the value of the corresponding game is easily determined.

Theorem 13.23 *Let G be a graph with $V(G) = \{1, \ldots, n\}$ and $E(G) = \{e_1, \ldots, e_m\}$. Let M be the $n \times m$ (0–1)-incidence matrix of G. Then the following assertions hold.*

(i) *If G is bipartite then $\nu(M) = \frac{1}{\nu(G)}$.*

(ii) *If G is the path then $\nu(M) = \frac{2}{n}$ if n is even, and $\frac{2}{n-1}$ if n is odd.*

(iii) *If G has a perfect matching then $\nu(M) = \frac{2}{n}$.*

(iv) *If G is Hamiltonian then* $v(M) = \frac{2}{n}$.
(v) *If G* $= K_n$ *then* $v(M) = \frac{2}{n}$.

Proof If G is bipartite, then by the König–Egervary theorem, $v(G) = \tau(G)$, and (i) follows by Lemma 13.22. Since a path is bipartite, (ii) follows from (i) and the fact that the matching number of a path on n vertices is $\frac{n}{2}$ if n is even and $\frac{n-1}{2}$ if n is odd.

If G has a perfect matching then $v(G) = \tau(G) = \frac{n}{2}$, and (iii) follows from (i).

To prove (iv), first suppose that G is the cycle on n vertices. Then $n = m$ and the strategies for Players I and II, which choose all pure strategies uniformly with probability $\frac{1}{n}$, are easily seen to be optimal. Thus, $v(M) = \frac{2}{n}$.

Suppose G is Hamiltonian. The value of M is at least equal to the value of the game corresponding to a Hamiltonian cycle in G and thus $v(M) \geq \frac{2}{n}$, in view of the preceding observation. If Player II chooses only the edges in the Hamiltonian cycle with equal probability, then against any pure strategy of Player I, Player II loses at most $\frac{2}{n}$. Therefore, (iv) is proved.

Finally, (v) follows since a complete graph is clearly Hamiltonian. □

Exercises

1. Let the matrix A be the direct sum of the matrices $A_1, \ldots, A_k$, that is,

$$A = \begin{bmatrix} A_1 & 0 & \cdots & 0 \\ 0 & A_2 & \cdots & 0 \\ \vdots & \vdots & \ddots & \vdots \\ 0 & 0 & \cdots & A_k \end{bmatrix}.$$

If $v(A_i) > 0$, $i = 1, \ldots, k$, then show that

$$v(A) = \left\{ \sum_{i=1}^{k} \frac{1}{v(A_i)} \right\}^{-1}.$$

Hence, determine the value of a square diagonal matrix.

2. Let G be a directed graph and let A be the skew matrix of G. Consider the matrix game A. Show that the dimension of the optimal strategy set and the number of essential strategies of a player are of the same parity.

3. Let G be a directed graph and let A be the skew matrix of G. Consider the matrix game A. Suppose every pure strategy is essential. Show that the dimension of the optimal strategy set equals $n - 1 - \text{rank } A$.

4. Let G be an acyclic directed graph with n vertices, m edges, $m \geq 2$, and let Q be the incidence matrix of G. Show that $v(Q) \geq \frac{2}{m(m-1)}$.

5. Consider the graph G:

Show that there are more than one optimal strategies for Player I in the corresponding incidence matrix game.

For an introduction to game theory, including matrix games, see [Ow82, Tij03]. Proofs of Theorems 13.4 and 13.5 can be found in [BKS50, GS50]. Relevant references for various sections are as follows: Sect. 12.2: [MQ06], Sect. 12.3: [FR92], Sect. 12.4: [BT97].

References and Further Reading

[BT97] Bapat, R.B., Tijs, S.: Incidence matrix games. In Game Theoretical Applications to Economics and Operations Research (Bangalore, 1996), Theory Decis. Lib. Ser. C, Game Theory Math. Program. Oper. Res., vol. 18, pp. 9–16. Kluwer Academic Publishers, Boston (1997)

[BKS50] Bohnenblust, H.F., Karlin, S., Shapley, L.S.: Solutions of discrete two-person games. In: Kuhn, H.W., Tucker, A.W. (eds.) Contrbutions to the Theory of Games, vol. 1, pp. 51–72. Princeton University Press, Princeton (1950)

[FR92] Fisher, D.C., Ryan, J.: Optimal strategies for a generalized "scissors, paper, and stone" game. Am. Math. Mon. **99**, 935–942 (1992)

[GS50] Gale, D., Sherman, S.: Solutions of finite two-person games. In: Kuhn, H.W., Tucker, A.W. (eds.) Contributions to the Theory of Games, vol. 1, pp. 37–49. Princeton University Press, Princeton (1950)

[MQ06] Michael, T.S., Quint, T.: Optimal strategies for node selection games on oriented graphs: skew matrices and symmetric games. Linear Algebra Appl. **412**, 77–92 (2006)

[Ow82] Owen, G.: Game Theory, 2nd edn. Academic Press, New York (1982)

[Tij03] Tijs, S.: Introduction to Game Theory. Texts and Readings in Mathematics. vol. 23, Hindustan Book Agency, New Delhi (2003)

Hints and Solutions to Selected Exercises

Chapter 1

1. $Ax = 0$ clearly implies $A'Ax = 0$. Conversely, if $A'Ax = 0$ then $x'A'Ax = 0$, which implies $(Ax)'Ax = 0$, and hence $Ax = 0$.
4. If $G = A^+$ then the two equations are easily verified. Conversely, suppose $A'AG = A'$ and $G'GA = G'$. Since rank $A'A = $ rank A, we may write $A = XA'A$ for some X. Then $A = XA'A = XA'AGA = AGA$. Also, $A'AG = A'$ implies $G'A' = G'A'AG = (AG)'AG$, which is symmetric. Similarly, using $G'GA = G'$, we may conclude that $GAG = G$ and that GA is symmetric.
5. $A = xy'$ for some column vectors x and y. First determine x^+ and y^+. $\alpha = (\text{trace } A'A)^{-1}$.

Chapter 2

2. Suppose $y_i = 1$, $y_j = -1$ and $y_k = 0$, $k \neq i$, $k \neq j$. Consider an (ij)-path $\mathscr{P}$. Let x be a vector with its coordinates indexed by $E(G)$. Set $x_k = 0$ if e_k is not in $\mathscr{P}$. Otherwise, set $e_k = 1$ or $e_k = -1$ according as e_k is directed in the same way as, or in the opposite way to, $\mathscr{P}$, respectively. Verify that $Qx = y$.
3. $Q^+ = Q'(QQ')^+$. Note that QQ' has a simple structure.
4. If G is not bipartite, then it has an odd cycle. Consider the submatrix of M corresponding to the cycle.
5. This is the well-known Frobenius–König theorem. Let G be the bipartite graph with bipartition (X, Y), where $X = Y = \{1, \ldots, n\}$, and i and j are adjacent if and only if $a_{ij} = 1$. Condition (i) is equivalent to $\nu(G) < n$. Use Theorem 2.22.

© Springer-Verlag London 2014
R.B. Bapat, *Graphs and Matrices*, Universitext,
DOI 10.1007/978-1-4471-6569-9

Chapter 3

1. The characteristic polynomial of either graph is $\lambda^6 - 7\lambda^4 - 4\lambda^3 + 7\lambda^2 + 4\lambda - 1$.
5. $(K_n) = 2(n-1), (K_{mn}) = 2\sqrt{mn}$.
6. Use Lemma 3.25.
7. Use the previous exercise to find the eigenvalues of the two graphs.
8. Note that $A(G_1) = \begin{bmatrix} 0 & 1 \\ 1 & 0 \end{bmatrix} \otimes A$ and $A(G_2) = \begin{bmatrix} 1 & 1 \\ 1 & 1 \end{bmatrix} \otimes A$, respectively. Use Lemma 3.25.
9. Let n be pendant and suppose it is adjacent to $n-1$. Assume the result for $T \setminus \{n\}$ and proceed by induction on n.

Chapter 4

1. A repeated application of Laplace expansion shows that $\det(L + J)$ is equal to the sum of $\det L$ and the sum of all cofactors of L. (Also see Lemma 9.3.) Use Theorem 4.8.
2. Let $|V(G)| = n$ and $|V(H)| = m$. Then $L(G \times H) = L(G) \otimes I_m + I_n \otimes L(H)$. If $\lambda_1, \ldots, \lambda_n$ and $\mu_1, \ldots, \mu_m$ are the eigenvalues of $L(G)$ and $L(H)$, respectively, then the eigenvalues of $L(G \times H)$ are $\lambda_i + \mu_j$; $i = 1, \ldots, n$; $j = 1, \ldots, m$.
3. Use Theorem 4.11 and the arithmetic mean-geometric mean inequality.
4. Use Theorem 4.13.
5. Let (X, Y) be a bipartition of T. Make all edges oriented from X to Y. The result holds for any bipartite graph.
6. For the first part, verify that $(A^+)'$ satisfies the definition of the Moore–Penrose inverse of A'. Then, for the second part note, using the first part, that

$$AA'(A')^+ A^+ AA' = AA'(A^+)' A^+ AA'.$$

Since the column space of A' is the same as that of A^+, it follows that $A^+ AA' = A'$. Substituting in the previous equation and using the first part shows that $(A')^+ A^+$ is a g-inverse of AA'. The other Moore–Penrose conditions are proved similarly.

Chapter 5

1. Note that $\begin{bmatrix} B \\ C \end{bmatrix} = \begin{bmatrix} I & B_f \\ -B'_f & I \end{bmatrix}$. By the Schur complement formula the determinant of $\begin{bmatrix} B \\ C \end{bmatrix}$ is seen to be nonzero.

2. Let B be the fundamental cut matrix. There exists an $(n-1) \times (n-1)$ nonsingular matrix Z such that $X' = ZB$. Use the fact that B is totally unimodular.
3. The proof is similar to that of Theorem 5.13.
4. First show that $\det BB'[E(T_1)|E(T_1)]$ is the number of spanning trees of G containing T_1 as a subtree. Use this observation and Theorem 4.7.

Chapter 6

1. Let A be the adjacency matrix of G and suppose $u \geq 0$ satisfies $Au = \mu u$. There exists $x > 0$ such that $Ax = \rho(G)x$. Consider $u'Ax$.
2. The Perron eigenvalue of a cycle and of $K_{1,4}$ is 2.
3. Use Corollary 6.16.
4. If G is strongly regular with parameters (n, k, a, c) then G^c is strongly regular with parameters (n, k_1, a_1, c_1), where $k_1 = n - k - 1$, $a_1 = n - 2k - 2 + c$ and $c_1 = n - 2k + a$.
5. For the first part use Theorem 6.27.
6. Let $\lambda_1, \ldots, \lambda_n$ be the eigenvalues of A. Since A is nonsingular, the eigenvalues are nonzero. By the arithmetic mean-geometric mean inequality,

$$\sum_{i=1}^{n} |\lambda_i| \geq n \prod_{i=1}^{n} |\lambda_i|^{\frac{1}{n}} = n |\det A|^{\frac{1}{n}} \geq n.$$

Chapter 7

2. Let vertex 1 be pendant, with neighbour 2. Then the adjacency matrix has the form

$$A = \begin{pmatrix} 0 & 1 & 0 & \cdots & 0 \\ 1 & 0 & \cdots & \cdots & \cdot \\ 0 & \cdot & & & \\ 0 & \vdots & & B & \\ 0 & \cdot & & & \end{pmatrix}.$$

It follows, using the determinantal definition of rank, that rank A = rank $B + 2$.
4. $(-1)^{r+s-1}((r-1)(s-2) + (r-2)(s-1))$.
5. $(r-1)(s-1) - (r-2)(s-2)$.

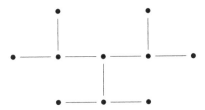

6.

7. If B is obtained from A by changing some 1's into -1's, then the parity of the determinant does not change. We may obtain B from A by changing some 1's into -1's, so that B is skew-symmetric. Now use the fact that a skew-symmetric matrix of odd order is singular.

8. If $n = 2$, the result holds. Let $n > 2$. If all entries are even, then $\det A \equiv 0 \bmod 4$. So let $P = \begin{pmatrix} x_1 & y \\ y & x_2 \end{pmatrix}$ be a principal submatrix, where x_1, x_2 are even and y is odd. Then $P^{-1} \equiv \begin{pmatrix} -x_1 & y \\ y & -x_2 \end{pmatrix}$ mod 4. Without loss of generality, $A = \begin{pmatrix} P & B \\ B' & C \end{pmatrix}$. Use the formula $\det A = (\det P) \det(C - B'P^{-1}B)$ and proceed by induction on n.

9. Use Theorem 3.8. Note that Exercise 7 is an easy consequence.

10. Use the preceding two Exercises.

11. For the first part, use induction on the number of vertices.

Chapter 8

1. The Laplacian L of $K_{1,n-1}$ has I_{n-1} as a principal submatrix. Therefore, the rank of $L - I_{n-1}$ is 2 and hence its nullity is $n - 2$. Thus, 1 is an eigenvalue of L with multiplicity $n - 2$. Clearly, 0 is an eigenvalue. The remaining eigenvalue, easily found using the trace, is n.

4. Use a symmetry argument.

5. The first part is an easy consequence of Theorem 8.16. For the second part, using the fact that Q_n is the n-fold Cartesian product of Q_2, show that the algebraic connectivity of Q_n is 2. Also see the remark following Corollary 8.18.

6. Let $f : V(G) \to \{0, 1, -1\}$ be defined by setting it equal to 0 on V_1, 1 on V_2 and -1 on V_3. Use the inequality $f'Lf \geq \mu f'f$ where L is the Laplacian. For a generalization and an application to "competitive learning process", see [3].

7. This is an easy consequence of Theorem 8.20.

8. Use Exercise 7.

9. Let $f_i = (n + 1) - 2i$, $i = 1, \ldots, n$. Note that $f'\mathbf{1} = 0$. Use (8.17).

Chapter 9

1. Follow an argument similar to that in the proof of Theorem 9.2.

3. For $\alpha \neq 0$ evaluate $\begin{vmatrix} -D & 1 \\ 1' & \frac{1}{\alpha} \end{vmatrix}$ two different ways.

5. Suppose $(D^{-1} - S)x = 0$ for some vector x. Premultiply this equation by $1'$ and use the formula for D^{-1} given in Theorem 9.9 to conclude $\tau'x = 0$ and hence that $(-\frac{1}{2}L - S)x = 0$, where L is the Laplacian of T. Then $x'(-\frac{1}{2}L - S)x = 0$, and since $\frac{1}{2}L + S$ is positive semidefinite, conclude that $x = 0$.

7. For any $i, j, k \in V(T)$, $d_{ij} = d_{ik} + d_{kj}$ mod 2.

8. Use (9.26) and that L^+, being positive semidefinite, is a Gram matrix, that is, there exist points $x^1, \ldots, x^n$ in $\mathbb{R}^n$ such that $\ell_{ij}^+ = (x^i)'x^j$, $i, j = 1, \ldots, n$.

10. Observe that I_k is a principal submatrix of the Laplacian matrix of T. Use interlacing and then apply Theorem 9.16.

Chapter 10

2. The resistance distance between any two vertices of the cycle is easily found by series-parallel reduction. Lemma 10.9 and a symmetry argument may also be used.

3. First prove the result when there is a cycle containing i and j. Then use the fact that if there are two (ij)-paths then there is an (ij)-path that meets a cycle.

4. By Theorem 10.12, if x is an $n \times 1$ vector orthogonal to τ, then $x'Rx \leq 0$.

6. Use (10.3) and Theorem 4.7.

7. There is a one-to-one correspondence between the spanning trees of G not containing the edge e_k and the spanning trees of G^* containing e'_k. Use the equation

$$\frac{\chi'(G)}{\chi(G)} + \frac{\chi(G) - \chi'(G)}{\chi(G)} = 1$$

and the previous exercise.

9. Use Theorem 10.12, the multilinearity of the determinant and the fact that each cofactor of L equals $\chi(G)$.

10. Assume that n is a pendant vertex and that the formula holds for $T \setminus \{n\}$. Use induction on the number of vertices.

Chapter 11

1. Use the recursive definition of a threshold graph and induction.

2. We may encode a threshold graph by a binary sequence $b_1, \ldots, b_n$, with $b_1 = 1$. In the recursive procedure to obtain the graph we add an isolated vertex if $b_i = 0$, and a dominating vertex if $b_i = 1$.

3. Use the recursive definition of a threshold graph and induction.

4. Use the recursive definition of a cograph and the fact that the union of two Laplacian integral graphs is Laplacian integral and the complement of a Laplacian integral graph is Laplacian integral.

5. Whether a graph G is a cograph or not can be checked recursively. Take the complement of G. Then it should split into connected components, each of which must be a cograph. Thus, if we take components of G^c and repeat the procedure of taking complements, we must end up with isolated vertices if the graph is a cograph. The presence of P_4 will not lead to this situation since P_4 is self-complementary. Incidentally, it is known that the property of not containing a P_4 as an induced subgraph characterizes cographs.

6. The eigenvalues of $L(G)$ are: n with multiplicity $|V_1|$, $|V_1|$ with multiplicity $|V_2| - 1$; and 0. The number of spanning trees in $K_m \backslash G$ is

$$m^{m-n-1}(m - |V_1|)^{|V_2|-1}(m - n)^{|V_2|}.$$

7. The eigenvalues of $L(G)$ are given by: $2r + 2$, $2r + 1$, $r + 2$ with multiplicity r, $r + 1$ with multiplicity $2r - 2$, r with multiplicity r, 1 and 0.

8. The eigenvalues of $L(K_n \times K_2)$ are: $n + 2$ with multiplicity $n - 1$; n with multiplicity $n - 1$; 2; and 0. (see [2].)

Chapter 12

1. A graph is rank k completable if and only if its bipartite complement has a matching of size k.

2. Use the definition of chordal graph. If G is a split graph then so is G^c.

5. Let G be the graph with $V(G) = \{1, \ldots, 7\}$ and with $i \sim j$ if and only if $a_{ij} \neq 0$. Then G is chordal and a perfect elimination ordering for G is given by $1, 2, 4, 5, 6, 7, 3$. Perform Gaussian elimination using pivots according to this ordering. So, first subtract a suitable multiple of a first row from the other rows to reduce all entries in the first column to zero except the $(1, 1)$-entry. Then subtract a suitable multiple of the first column from the remaining columns to reduce all entries in the first row to zeros, except the $(1, 1)$-entry. Repeat the process with the second row and column, then with the fourth row and column, and so on. In the process, no zero entry will be changed to a nonzero entry.

6. Use the Jacobi identity.

Chapter 13

2. Use Theorem 13.4 and the fact that the rank of a skew-symmetric matrix is even.
3. The optimal strategy set comprises the vectors in $\mathcal{P}_n$ that are in the null space of A.
4. It is sufficient to show that $\sum_v \rho(v) \leq \frac{m(m-1)}{2}$. Let u be a source. Assume the result for $G \setminus \{u\}$ and proceed by induction on the number of vertices.
5. $[\frac{1}{3}, \frac{1}{3}, \frac{1}{3}, 0]'$ and $\frac{1}{4}\mathbf{1}'$ are both optimal for Player I.

Bibliography

[Akbari07]   Akbari, S., Kirkland, S.J.: On unimodular graphs. Linear Algebra Appl. **421**, 3–15 (2007)

[Alon85]   Alon, N., Milman, V.D.: λ_1, isoperimetric inequalities for graphs and supercon-centrators. J. Comb. Theory, Ser. B **38**, 73–88 (1985)

[Anderson85]   Anderson, W.N., Morley, T.D.: Eigenvalues of the Laplacian of a graph. Linear Multilinear Algebra **18**(2), 141–145 (1985)

[Bai11]   Bai, H.: The Grone-Merris conjecture. Trans. Amer. Math. Soc. **363**(8), 4463–4474 (2011)

[Balakrishnan04] Balakrishnan, R.: The energy of a graph. Linear Algebra Appl. **387**, 287–295 (2004)

[Bapat97]   Bapat, R.B.: Moore-Penrose inverse of the incidence matrix of a tree. Linear Multilinear Algebra **42**, 159–167 (1997)

[Bapat99]   Bapat, R.B.: Resistance distance in graphs. Math. Stud. **68**, 87–98 (1999)

[Bapat2000]   Bapat, R.B.: Linear Algebra and Linear Models, 2nd edn. Hindustan Book Agency, New Delhi (2000)

[Bapat04]   Bapat, R.B.: Resistance matrix of a weighted graph. MATCH Commun. Math. Comput. Chem. **50**, 73–82 (2004)

[Bapat11]   Bapat, R.B.: A note on singular line graphs. Bull. Kerala Math. Assoc. **8**(2), 207–209 (2011)

[Bapat08]   Bapat, R.B., Gupta, S.: Resistance matrices of blocks in a graph. AKCE Int. J. Graphs Combinatorics **5**(1), 35–45 (2008)

[Bapat05]   Bapat, R., Kirkland, S.J., Neumann, M.: On distance matrices and Laplacians. Linear Algebra Appl. **401**, 193–209 (2005)

[Bapat98]   Bapat, R.B., Pati, S.: Algebraic connectivity and the characteristic set of a graph. Linear Multilinear Algebra **45**, 247–273 (1998)

[Bapat02]   Bapat, R.B., Pati, S.: Path matrices of a tree. Journal of Mathematical Sciences, New Series (Delhi) vol.1, pp. 46–52. (2002)

[BapatPati04]   Bapat, R.B., Pati, S.: Energy of a graph is never an odd integer. Bull. Kerala Math. Assoc. **1**(2), 129–132 (2004)

[Bapat 97]   Bapat, R.B., Raghavan, T.E.S.: Nonnegative Matrices and Applications, Encyclopedia of Mathematics and Its Applications, 64. Cambridge University Press, Cambridge (1997)

[Bapat96]   Bapat, R.B., Tijs, S.: Incidence matrix games. In Game Theoretical Applications to Economics and Operations Research (Bangalore, 1996), Theory Decis. Lib.

© Springer-Verlag London 2014
R.B. Bapat, *Graphs and Matrices*, Universitext,
DOI 10.1007/978-1-4471-6569-9

Ser. C, Game Theory Math. Program. Oper. Res., vol. 18, pp. 9–16. Kluwer Academic Publishers, Boston (1997)

[Barik06] Barik, S., Neumann, M., Pati, S.: On nonsingular trees and a reciprocal eigenvalue property. Linear Multilinear Algebra **54**(6), 453–465 (2006)

[Ben03] Ben-Israel, A., Greville, T.N.E.: Generalized Inverses. Theory and Applications, 2nd edn. Springer, New York (2003)

[Berman94] Berman, A., Plemmons, R.J.: Nonnegative Matrices in the Mathematical Sciences, Classics in Applied Mathematics, 9. SIAM, Philadelphia (1994)

[Berman03] Berman, A., Shaked-Monderer, N.: Completely Positive Matrices. World Scientific, Singapore (2003)

[Bevis81] Bevis, J.H., Hall, F.J., Katz, I.J.: Integer generalized inverses of incidence matrices. Linear Algebra Appl. **39**, 247–258 (1981)

[Biggs93] Biggs, N.: Algebraic Graph Theory, 2nd edn. Cambridge University Press, Cambridge (1993)

[Biggs92] Biggs, N.L., Brightwell, G.R., Tsoubelis, D.: Theoretical and practical studies of a competitive learning process. Netw. Comput. Neural Syst. **3**(3), 285–301 (1992)

[Bohnenblust50] Bohnenblust, H.F., Karlin, S., Shapley, L.S.: Solutions of discrete two-person games. In: Kuhn, H.W., Tucker, A.W. (eds.) Contrbutions to the Theory of Games, vol. 1, pp. 51–72. Princeton University Press, Princeton (1950)

[Bollob98] Bollobás, B.: Modern Graph Theory. Springer, New York (1998)

[Bondy08] Bondy, J.A., Murty, U.S.R.: Graph Theory, Graduate Texts in Mathematics, 244. Springer, New York (2008)

[Buckley88] Buckley, F., Doty, L.L., Harary, F.: On graphs with signed inverses. Networks **18**(3), 151–157 (1988)

[Cameron78] Cameron, P.J.: Strongly regular graphs. In: Beineke, L.W., Wilson, R.J. (eds.) Selected Topics in Graph Theory, pp. 337–360. Academic Press, New York (1978)

[Campbell79] Campbell, S.L., Meyer, C.D.: Generalized Inverses of Linear Transformation. Pitman, Pitman (1979)

[Cioab09] Cioabă, S.M., Ram Murty, M.: A First course in Graph Theory and Combinatorics, Texts and Readings in Mathematics 55. Hindustan Book Agency, New Delhi (2009)

[Cvetkovi95] Cvetković, D.M., Doob, M., Sachs, H.: Spectra of Graphs, Theory and Applications, 3rd edn. Johann Ambrosius Barth, Heidelberg (1995)

[Das03] Das, K.Ch.: An improved upper bound for Laplacian graph eigenvalues. Linear Algebra Appl. **368**, 269–278 (2003)

[Deo74] Deo, N.: Graph Theory with Applications to Engineering and Computer Science. Prentice-Hall Inc, New Jersey (1974)

[Doyle84] Doyle, P.G., Snell, J.L.: Random Walks and Electrical Networks. Mathematical Association of America, Washington (1984)

[Faria85] Faria, I.: Permanental roots and the star degree of a graph. Linear Algebra Appl. **64**, 255–265 (1985)

[Fiedler73] Fiedler, M.: Algebraic connectivity of graphs. Czech. Math. J. **23**(98), 298–305 (1973)

[Fiedler75] Fiedler, M.: Eigenvalues of acyclic matrices. Czech. Math. J. **25**(100), 607–618 (1975)

[Fisher92] Fisher, D.C., Ryan, J.: Optimal strategies for a generalized "scissors, paper, and stone" game. Am. Math. Mon. **99**, 935–942 (1992)

[Gale50] Gale, D., Sherman, S.: Solutions of finite two-person games. In: Kuhn, H.W., Tucker, A.W. (eds.) Contributions to the Theory of Games, vol. 1, pp. 37–49. Princeton University, Princeton (1950)

[Ebrahim12] Ghorbani, E.: Spanning trees and line graph eigenvalues. arXiv:1201.3221v1 (2012)

[Ebrahim13]   Ghorbani, E.: Spanning trees and line graph eigenvalues. arXiv:1201.3221v3 (2013)

[Godsil93]   Godsil, C.D.: Algebraic Comb. Chapman and Hall Inc, New York (1993)

[Godsil01]   Godsil, C., Royle, G.: Algebraic Graph Theory, Graduate Texts in Mathematics, 207. Springer, New York (2001)

[Golumbic80]   Golumbic, M.: Algorithmic Graph Theory and Perfect Graphs. Academic Press, New York (1980)

[Graham77]   Graham, R.L., Hoffman, A.J., Hosoya, H.: On the distance matrix of a directed graph. J. Comb. Theory **1**, 85–88 (1977)

[Graham78]   Graham, R.L., Lovász, L.: Distance matrix polynomials of trees. Adv. Math. **29**(1), 60–88 (1978)

[Graham71]   Graham, R.L., Pollak, H.O.: On the addressing problem for loop switching. Bell. Syst. Tech. J. **50**, 2495–2519 (1971)

[Grone84]   Grone, R., Johnson, C.R., Sá, E.M., Wolkowitz, H.: Positive definite completions of partial hermitian matrices. Linear Algebra Appl. **58**, 109–124 (1984)

[Grone94]   Grone, R., Merris, R.: The Laplacian spectrum of a graph II. SIAM J. Discrete Math. **7**(2), 221–229 (1994)

[Grone08]   Grone, R., Merris, R.: Indecomposable Laplacian integral graphs. Linear Algebra Appl. **428**, 1565–1570 (2008)

[Grone90]   Grone, R., Merris, R., Sunder, V.S.: The Laplacian spectrum of a graph. SIAM J. Matrix Anal. Appl. **11**, 218–238 (1990)

[Grossman95]   Grossman, J.W., Kulkarni, D., Schochetman, I.E.: On the minors of an incidence matrix and its Smith normal form. Linear Algebra Appl. **218**, 213–224 (1995)

[Ivan01]   Gutman, Ivan, Sciriha, Irene: On the nullity of line graphs of trees. Discrete Math. **232**, 35–45 (2001)

[Horn85]   Horn, R.A., Johnson, C.R.: Matrix Anal. Cambridge University Press, Cambridge (1985)

[Ijiri65]   Ijiri, Y.: On the generalized inverse of an incidence matrix. J. Soc. Ind. Appl. Math. **13**(3), 827–836 (1965)

[Indulal06]   Indulal, G., Vijayakumar, A.: On a pair of equienergetic graphs. MATCH Commun. Math. Comput. Chem. **55**(1), 83–90 (2006)

[Klein93]   Klein, D.J., Randić, M.: Resistance distance. J. Math. Chem. **12**, 81–95 (1993)

[Koolen01]   Koolen, J.H., Moulton, V.: Maximal energy graphs. Adv. Appl. Math. **26**, 47–52 (2001)

[Lov86]   Lovász, L., Plummer, M.D.: Matching Theory, Annals of Discrete Mathematics, 29. North-Holland, Amsterdam (1986)

[Mahadev95]   Mahadev, N.V.R., Peled, U.N.: Threshold Graphs and Related Topics, Annals of Discrete Mathematics, 54. North-Holland, Amsterdam (1995)

[Marshall79]   Marshall, A.W., Olkin, I.: Inequalities: Theory of Majorization and Its Applications, Mathematics in Science and Engineering, 143. Academic Press, New York, London (1979)

[Merris87]   Merris, R.: Characteristic vertices of trees. Linear Multilinear Algebra **22**, 115–131 (1987)

[Merris89]   Merris, R.: An edge version of the matrix-tree theorem and the Wiener index. Linear Multilinear Algebra **25**, 291–296 (1989)

[Merris90]   Merris, R.: The distance spectrum of a tree. J. Graph Theory **3**(14), 365–369 (1990)

[Merris94]   Merris, R.: Degree maximal graphs are Laplacian integral. Linear Algebra Appl. **199**, 381–389 (1994)

[Michael06]   Michael, T.S., Quint, T.: Skew matrices and symmetric games: Optimal strategies for node selection games on oriented graphs. Linear Algebra Appl. **412**, 77–92 (2006)

[Moon95]   Moon, J.W.: On the adjoint of a matrix associated with trees. Linear Multilinear Algebra **39**, 191–194 (1995)

[Owen82]   Owen, G.: Game Theory, 2nd edn. Academic Press, New York (1982)
[Recski89]   Recski, A.: Matroid Theory and Its Applications in Electric Network Theory and
                in Statics, Algorithms and Combinatorics, 6. Springer, Berlin (1989)
[Schwenk78]   Schwenk, A.J., Wilson, R.J.: On the eigenvalues of a graph. In: Beineke, L.W.,
                Wilson, R.J. (eds.) Selected Topics in Graph Theory, pp. 307–336. Academic
                Press, New York (1978)
[Irene98]   Sciriha, Irene: On singular line graphs of trees. Congr. Numeratium **135**, 73–91
                (1998)
[So99]   So, W.: Rank one perturbation and its application to the Laplacian spectrum of
                a graph. Linear Multilinear Algebra **46**, 193–198 (1999)
[Stanley96]   Stanley, R.: A matrix for counting paths in acyclic digraphs. J. Comb. Theory,
                Ser. A **74**, 169–172 (1996)
[Tijs03]   Tijs, S.: Introduction to Game Theory, Texts and Readings in Mathematics, 23.
                Hindustan Book Agency, New Delhi (2003)
[West02]   West, D.: Introduction to Graph Theory, 2nd edn. Prentice-Hall, India (2002)

Index

© Springer-Verlag London 2014
R.B. Bapat, *Graphs and Matrices*, Universitext,
DOI 10.1007/978-1-4471-6569-9